Thermo-energetic Design of Machine Tools

Lecture Notes in Production Engineering

More information about this series at http://www.springer.com/series/10642

Knut Großmann
Editor

Thermo-energetic Design of Machine Tools

A Systemic Approach to Solve the Conflict Between Power Efficiency, Accuracy and Productivity Demonstrated at the Example of Machining Production

Springer

Editor
Knut Großmann
Institut für Werkzeugmaschinen
und Steuerungstechnik
TU Dresden
Dresden
Germany

ISSN 2194-0525 ISSN 2194-0533 (electronic)
ISBN 978-3-319-36546-6 ISBN 978-3-319-12625-8 (eBook)
DOI 10.1007/978-3-319-12625-8

Springer Cham Heidelberg New York Dordrecht London

Printed on acid-free paper

Springer is part of Springer Science+Business Media (www.springer.com)

Contents

1 **Introduction** 1
Knut Großmann and Gritt Ott

2 **Model-Based Representation of Thermo-energetic Effects in Cutting Tools and Part Clamping Devices** 13
Michael Bräunig, Ulrich Semmler, Gerhard Schmidt, Volker Wittstock and Matthias Putz

3 **Model and Method for the Determination and Distribution of Converted Energies in Milling Processes** 27
Matthias Brockmann, Fritz Klocke and Drazen Veselovac

4 **Energy Model for Grinding Processes** 35
Matthias Rasim, Fritz Klocke and Patrick Mattfeld

5 **Thermo-energetic Modelling of Fluid Power Systems** 49
Juliane Weber and Jürgen Weber

6 **Simulation of Pose- and Process-Dependent Machine Tool Models** 61
Marian Partzsch and Michael Beitelschmidt

7 **Thermo-Elastic Simulation of Entire Machine Tool** 69
Alexander Galant, Knut Großmann and Andreas Mühl

8 **Model Order Reduction for Thermo-Elastic Assembly Group Models** 85
Norman Lang, Jens Saak and Peter Benner

9 **High-Accuracy Thermo-Elastic Simulation on Massively Parallel Computer** 95
Andreas Naumann, Florian Stenger, Axel Voigt and Jörg Wensch

10 **Modelling of Thermal Interactions Between Environment and Machine Tool** 111
Reimund Neugebauer, Steffen Ihlenfeldt, Carsten Zwingenberger, Janine Glänzel and Carsten Richter

11 **Determination and Modelling of Heat Transfer Mechanisms Acting Among Machine Tool Components** 125
Sarah Vieler, Yona Frekers, Michael Burghold and Reinhold Kneer

12 **Investigation of Components and Assembly Groups** 135
Christian Brecher, Marcel Fey, Dorothea Haber and Kolja Bakarinow

13 **Adjustment of Uncertain Parameters in Thermal Models of Machine Tools** 145
Bernd Kauschinger, Klaus Kabitzsch and Steffen Schroeder

14 **Correction Algorithms and High-Dimensional Characteristic Diagrams** 159
Christian Naumann, Ilka Riedel, Ulrich Priber and Roland Herzog

15 **Correction Model of Load-Dependent Structural Deformations Based on Transfer Functions** 175
Christian Brecher, Marcel Fey and Matthias Wennemer

16 **Structural Model-Based Correction of Thermo-elastic Machine Tool Errors** 185
Xaver Thiem, Knut Großmann and Andreas Mühl

17 **Modelling and Design of Systems for Active Control of Temperature Distribution in Frame Subassemblies** 199
Welf-Guntram Drossel, André Bucht and Christoph Ohsenbrügge

18 **Structurally Integrated Sensors** 209
Christian Wenzel and Michel Klatte

19 **Thermo-Energetic Motor Optimisation** 223
Stefan Winkler and Ralf Werner

20 Technical and Economic Benchmarking Guideline for the Compensation and Correction of Thermally Induced Machine Tool Errors 233
Hajo Wiemer, Lars Neidhardt, Werner Esswein and Richard Braun

21 Experimental Analysis of the Thermo-Elastic Behaviour of Machine Tools by Means of Selective Thermography and Close-Range Photogrammetry 247
Knut Großmann, Jens Müller, Marcel Merx and Mirko Riedel

Abbreviations

ADS	Automation Device Specification
AMDiS	Adaptive MultiDimensional Simulations
AS	Assemblies
BC	Boundary Conditions
BPMN	Business Process Management and Notation
BT	Balanced Truncation
CFD	Computational Fluid Dynamics
CFRP	Carbon Fibre-Reinforced Plastics
CL	Cooling Lubricant
CRC/TR 96	Collaborative Research Centres/Transregio 96
CS	Coordinate System
DBS	Block-oriented Digital Simulation
DFG	German Research Foundation
DoF	Degrees of Freedom
DPA	Digital Photogrammetric Analysis
FE	Finite Elements
FEM	Finite Element Method
GMRES	Generalized Minimal Residual Method
HMI	Human–Machine Interface
HPC	High Performance Cutting
IR	Infrared Camera
IRKA	Iterative Rational Krylov Algorithm
IWM	Institute for Machine Tools and Control Technology
IWU	Fraunhofer Institute for Machine Tools and Forming Technology
LTI	Linear Time Invariant
LTV	Linear Time Varying
MOR	Model Order Reduction
MOR Sim	Model Order Reduction Simulation
MR fluids	Magnetorheological fluids
MT	Machine Tool

NC	Numerical Control
ODE	Ordinary Differential Equation
PCM	Phase-Change Materials
PDE	Partial Differential Equation
PMOR	Parametric Model Order Reduction
POD	Proper Orthogonal Decomposition
SEM	Scanning Electron Microscope
SGR	Smoothed Grid Regression
SLS	Switched Linear Systems
TCP	Tool Center Point
VDMA	Association of German Machine and Plant Engineering
WZL	Laboratory for Machine Tools and Production Engineering

Chapter 1
Introduction

Knut Großmann and Gritt Ott

Abstract The approach to the solution within the CRC/TR 96 financed by the German Research Foundation DFG aims at measures that will allow manufacturing accuracy to be maintained under thermally unstable conditions with increased productivity, without an additional demand for energy for tempering. The challenge of research in the CRC/TR 96 derives from the attempt to satisfy the conflicting goals of reducing energy consumption and increasing accuracy and productivity in machining. In the current research performed in 19 subprojects within the scope of the CRC/TR 96, correction and compensation solutions that influence the thermo-elastic machine tool behaviour efficiently and are oriented along the thermo-elastic functional chain are explored and implemented.

1.1 Problem

With the aim of higher profitability, cutting machines are faced with the challenge of achieving both higher productivity and accuracy requirements (Kinkel and Lay 2006). This trend is accompanied by an ever more pronounced demand for improved energy efficiency in production. As demonstrated in Fig. 1.1, it is impossible to trace productivity, accuracy and energy consumption individually

Following a simplified assumption that expresses productivity by the metal removal rate Q_w and energy consumption by the cutting power P_C, then the following simplified relationships

$$P_C = K \cdot Q_W \quad \text{with} \quad Q_W = A \cdot v_f \tag{1.1}$$

K. Großmann · G. Ott (✉)
Faculty of Mechanical Engineering, Institute for Machine Tools and Control Engineering, Technical University Dresden, Dresden, Germany
e-mail: gritt.ott@tu-dresden.de

K. Großmann
e-mail: knut.grossmann@tu-dresden.de

K. Großmann (ed.), *Thermo-energetic Design of Machine Tools*,
Lecture Notes in Production Engineering, DOI 10.1007/978-3-319-12625-8_1

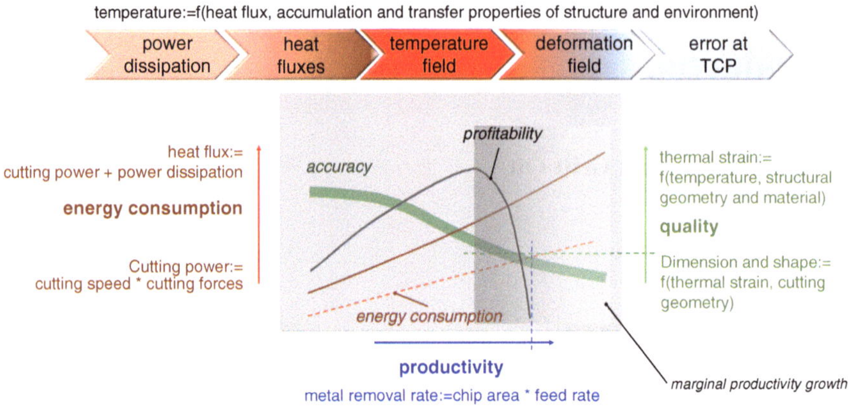

Fig. 1.1 Increase in productivity limited by increased heat input

$$K = k_C \cdot \frac{v_c}{v_f} \quad \text{and} \quad P_C = k_C \cdot A \cdot v_C$$

yield that energy consumption grows proportionally with productivity. In this equation, k_C is the specific cutting force, v_c the cutting speed, v_f the feed rate and A the chip area.

Since power dissipation P_v, in the form of heat flux is composed both of the power losses in the drives and moving contacts and that of the machining process, P_v is proportional to cutting power P_C and thus, in turn, productivity, which is quantified by the metal removal rate Q_w. Thus, productivity growth entails a greater amount of heat introduced in the machine tool structure. As a result, and according to the thermo-elastic functional chain, a temperature field is developed as a function of the thermal accumulation and transfer characteristics of the structure as well as the environmental conditions. Corresponding to

$$P_V = C_{therm} \cdot \frac{dT}{dt} + L \cdot \Delta T \tag{1.2}$$

the temperatures T and the temperature differences ΔT grow as a function of the power dissipation P_v introduced. In this equation, C_{therm} is the heat capacity (heat accumulation capacity) and L the thermal conductivity (heat transfer properties). The non-steadiness of the latter parameters grows due to increasing manufacturing dynamics. Due to thermal expansion immediately in the machine structure, this temperature field results in a thermally induced deformation field that depends on the structural geometry and material. At the same time, the driving, bearing and guidance bases of the machine tool's feed system are distorted and lead to kinematic errors at the TCP, which, in turn, depending on their location in the workspace and the manufacturing geometry, are represented as dimensional and form errors of the workpiece. Assuming, for a clear and simplified demonstration, that, for instance, the change in the structural length l

$$\Delta\, l = \alpha_{\text{therm}} \cdot l \cdot \Delta T \tag{1.3}$$

causes a kinematic error of $\Delta X = \Delta l$ directly situated in the direction of the workpiece dimension W due to the temperature difference ΔT and the thermal linear expansion coefficient α_{therm}, then we obtain for the accuracy

$$G_W = \frac{W}{\Delta X} \quad \text{with} \quad \Delta W = \Delta X, \tag{1.4}$$

when the accuracy is understood as the inverse of the relative deviation $\Delta W/W$. This means, assuming the strong simplifications mentioned above, that along the chain of proportionalities,

$$Q_W \sim P_C \sim P_V \sim \Delta T \sim \Delta l \sim \Delta X \sim \frac{1}{G_W} \tag{1.5}$$

one can detect a hyperbolic decrease in accuracy with increasing productivity.

The basic relationship remains valid, although the real conditions are much more complex as a result of the spatial machine structures, kinematic mechanisms and workpiece geometries, as well as the influence of nonlinearities.

It is currently possible, on the one hand, to keep the errors due to geometric-kinematic and static impacts on the workpiece low as a result of optimised design and the use of high quality components; however, on the other hand, the thermo-elastically induced error ratio nevertheless ranges from 70 to 75 % (Brecher et al. 2014). From this finding, one must conclude that the increases in productivity thanks to higher machine power have reached their limits, if no other measures are taken.

The measures currently practiced aim at achieve a thermal state of inertia by introducing power dissipations as widely as homogeneously as possible or by guaranteeing boundary conditions to the greatest extent possible. These conventional measures, for instance, include "warming up" in the steady-state condition before starting intrinsic manufacturing, the continuous use of thermally stabilising hydraulic and metalworking fluid cycles in idle times as well, tempering of structural supporting areas of the machine tools by means of fluids to be re-cooled and air-conditioning the entire production areas.

1.2 Objectives and Project Structure of the CRC/TR 96

However, the approach to the solution within the CRC/TR 96 financed by the German Research Foundation DFG aims at measures that will allow manufacturing accuracy to be maintained under thermally unstable conditions with increased productivity, without an additional demand for energy for tempering. The challenge of research in the CRC/TR 96 derives from the attempt to satisfy the conflicting goals of reducing energy consumption and increasing accuracy and productivity in machining.

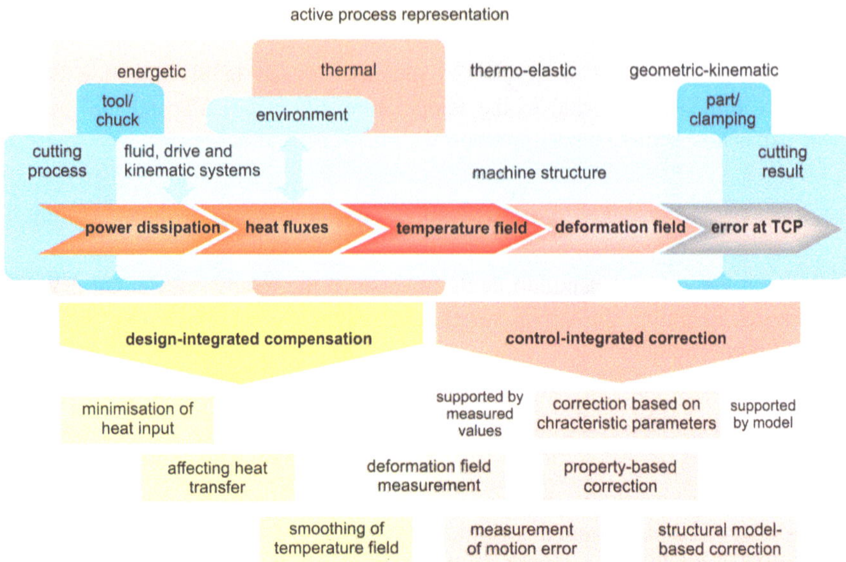

Fig. 1.2 Approach across the thermo-elastic functional chain

In the current research performed in 19 subprojects within the scope of the CRC/ TR 96, correction and compensation solutions that influence the thermo-elastic machine tool behaviour efficiently and are oriented along the thermo-elastic functional chain of Fig. 1.2 are explored and implemented. As part of this general objective, the following issues must be researched and engineered in an interdisciplinary setting and brought together into useful overall solutions:

1. Providing the modelling fundamentals to calculate the heat fluxes and the resulting thermo-elastic deformations in a comprehensive manner,
2. Mapping of the structural variability as a result of the relative movement inside the machine tool,
3. Providing the tools for an efficient adjustment of parameters that vary greatly in time and space by means of parameter identification methods as a prerequisite for correction and compensation solutions,
4. Engineering and demonstrating solutions to control-integrated correction of thermo-elastic errors by an inverse position setpoint compensation of the error at the TCP,
5. Engineering and demonstrating solutions based on the material properties to compensate for thermo-elastic effects through a homogeneous propagation of the temperature field, as well as reducing and smoothing the distribution of heat dissipated in supporting structures,

6. Developing metrological fundamentals to record the thermo-elastic errors in special structural areas of machine tools,
7. Engineering a methodological approach to simultaneous and complex evaluation of the CRC/TR 96 solutions, referring to their impact on product quality, production rate, energy consumption and machine tool costs.

Fundamental modelling and simulation technologies, as well as content related to the partial models for calculation and analysis of the thermo-elastic behaviour of assemblies and the active moving overall cutting machine system were engineered, coordinated and connected in the subprojects in the project area A "Active process machine tool model":

A01 Deformation of tool and clamping device (Chap. 2)
A02 Energy model for milling processes (Chap. 3)
A03 Energy model for grinding processes (Chap. 4)
A04 Thermo-fluid technology (Chap. 5)
A05 Simulation of active machine tool models (Chaps. 6 and 7)
A06 Model order reduction (Chap. 8)
A07 High-resolution thermo-elastic simulation (Chap. 9)

Project area B, entitled "Parameter setting and correction", and its corresponding subprojects are focussed on the development, investigation and application of alternative motion correction methods, as well as studies of parameter setting for the required models that represent thermo-elastic machine tool behaviour:

B01 Thermal interaction between environment and machine tool (Chap. 10)
B02 Heat transfer mechanisms between machine tool components (Chap. 11)
B03 Investigations of components and assemblies (Chap. 12)
B04 Object-related parameter identification (Chap. 13)
B05 Performance data based correction (Chap. 14)
B06 Property model based correction (Chap. 15)
B07 Structure model based correction (Chap. 16)

Finally, project area C, "Component and machine tool design and—evaluation", brings together the subprojects focussed on design that are required to design compensation solutions. Activities to implement, demonstrate and benchmark the correction, compensation and metrological solutions in a manner that covers the working areas and subjects, were coordinated:

C02 Controlled heat flux (Chap. 17)
C03 Structurally integrated sensors (Chap. 18)
C04 Thermo-energetic motor optimisation (Chap. 19)
C05 Benchmarking (Chap. 20)
C06 Test bed (Chap. 21)

1.3 Design-Integrated Compensation and Control-Integrated Correction

Design-integrated compensation affects a priori the power dissipation values transferred into heat and the formation of the temperature field. The objective is to represent heat input as a continuous process in case of transient load situations and to achieve a homogeneous distribution of the resulting temperature field. First, the basic mechanisms of latent heat accumulation in phase-change materials (PCM)/ metal foam, as well as heat bridges switched by means of magneto-rheological fluids, are investigated. These relationships are explored using primary beam and plate structures. (Chap. 17)

In control-integrated correction methods, the thermally induced kinematic errors that occur at the TCP have to be smoothed. As a prerequisite, they need active or updated process information about the variables to be corrected by means of feed drives or additional control axes. This information can be obtained from the models, on the one hand, and from direct deformation measurement, on the other hand.

The demand for a complete model that is effective for the active process refers not only to the thermal loads, but also to the structural changes caused by the assemblies' relative motions. To map the machine tool effectively for the active process, partial models of assemblies and components are brought together. (Chap. 7). They include the operation-dependent losses and heat flux distributions in the electrical drives (Chap. 19) and the fluid-technical systems (Chap. 5), the bearings' and guidance components' frictional losses depending on velocity and load (Chap. 12), as well as those heat quantities dissipating into the machine tool, which come from the technological parameters (Chaps. 3 and 4), and the distribution of the dissipated heat content to tool, workpiece and chip in the cutting process (Chap. 2). The resulting FE models are sophisticated and require lengthy computation times. Consequently, to make them feasible, they have to be reduced to a time-saving calculation model by means of innovative algorithms for model order reduction (Chap. 8). This reduction is necessary to run the control-integrated machine tool correction in real time mode. The models of reduced order were verified by simulation (Chap. 7).

The activities within the CRC/TR 96 investigate three alternative methods to correct the thermally induced kinematic errors in the control in real time.

- The correlation-based correction model (Chap. 14) explores the relationships between selected discrete temperature values from the machine structure and associated TCP displacements.
- Modelling for property model based correction (Chap. 15) is based on the response between the process-typical load parameters, such as the mechanical load profile and the thermo-elastically induced TCP displacements that occur.
- Structural model-based correction (Chap. 16) relies on the sufficiently reliable calculation of the temperature distribution to be run in real time and the resulting TCP displacement as a function of the effective process guidance via a structural

model. This model is generated from the order-reduced FE representations of the structural assemblies, as well as the heat source representations of the process and the drive, bearing and guidance components, synchronised to the process and updated in terms of parameters and structure.

1.4 Metrological Analysis and Benchmarking

The deformation measurement activities are performed with two aims. On the one hand, a structure-integrated sensor system (Chap. 18) is developed, which first directly captures local structural deformations, shown for the example of a machine bed. On the other hand, the opportunities for direct high-resolution TCP measurement by combining close range photogrammetry and selective thermography are investigated (Chap. 21).

To guarantee the required model quality for the correction methods, various fundamental investigations are performed. Furthermore, highly efficient and feasible strategies to assist modelling are engineered. A rapid, comprehensive and low-cost method is developed for the metrological adjustment of the parameters that are difficultly to find, dependent on each individual object and varying over the operation period, with the model (Chap. 13). The technical content of the FE models is also enriched by a more precisely determined contact heat transfer coefficient in joint components (Chap. 11), and the representation of the interface between the environment and the machine tool based on the FE method (Chap. 10).

An adequate benchmarking tool is created to benchmark all solution methods found within the scope of the CRC/TR 96 for the correction and/or compensation of thermally induced machining errors. The tool compares different consequences related to achievable accuracy and the production rate, as well the endeavours required for their generation and application (Chap. 20).

1.5 Test Beds and Demonstrators

The algorithms and strategies for modelling and parameter setting, correction- and compensation solutions, as well as for benchmarking, which are developed in the subprojects, are also implemented, tested and evaluated (benchmarked) on several test stands and demonstrators.

Table 1.1 shows the individual test beds and demonstrators and their locations in overview.

Through several subprojects, research was conducted on the assemblies and demonstrators introduced below.

The column of the Auerbach ACW630 belonged to a prototype-like machining centre by the firm Auerbach Maschinenfabrik GmbH. (Chaps. 10 and 14)

Table 1.1 Demonstrators

	Aachen	Chemnitz	Dresden
Test beds	Test beds for studies at ball screws, guidance systems and main spindles	Test bed equipped with measuring instruments: tool chuck and part clamping system in an air-conditioned room	Motor spindle sleeve bearing test bed
Analysed assemblies		Column of the Auerbach ACW630	Column assembly
Demonstrators	DMU 40eVo linear	Experimental machine tool HEC800	MiniHex test bed MAX

The column assembly was equipped with individually switchable heating films in the area of the z guidances that are insulated against the environment. This makes it possible to simulate in experiment a spindle head's movement, which is in reality connected with "non-steady" heat dissipation in the area of the guidances due to friction, in experiment (Fig. 1.3).

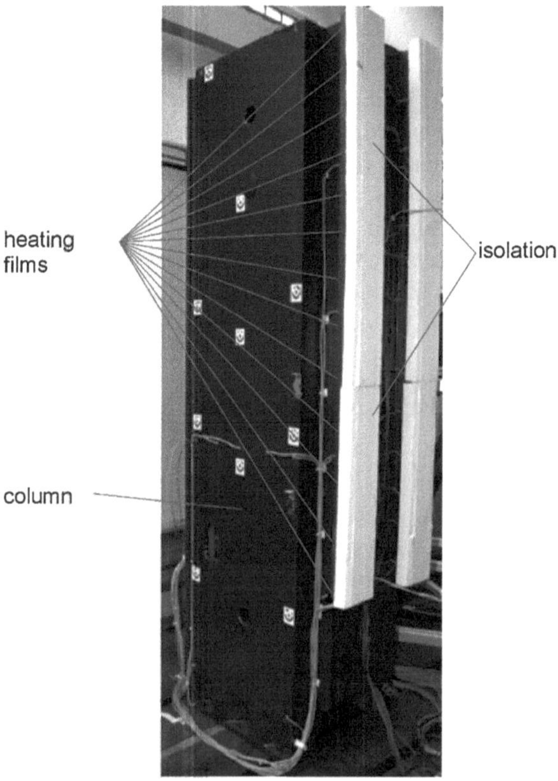

Fig. 1.3 Demonstrator of a milling machine column at the Dresden location

This setup is designed to test and verify the following solutions in experiment:

1. Column-spindle head motion—simulation of a thermal network model of variable structure which is correspondingly reduced in terms of model order (Chap. 6)
2. Testing the measurement methods developed for photogrammetric deformation measurement and testing the connection of this deformation measurement by means of thermographic temperature field measurement for selective thermal imaging (Chap. 21)
3. Comparison of alternative modelling methods to create reduced models with variable structure (Chap. 8) versus unreduced structurally variable (Chap. 9) thermo-elastic models for the temperature values really measured on the assembly as a function of time

In Dresden, the test bed MAX was built up; due to its lightweight construction based on aluminium structures, low heat capacities and high thermal coefficients of expansion are to be expected. Furthermore, high locally variable heat inputs are to be expected from structurally integrated direct linear drives run without cooling. The test bed embodies a three-axis Cartesian kinematic mechanism, in which all motions are carried out by the workpiece. In the cross slide, linear direct drives free of iron are used to generate the motion in X- and Y-axes. Correction motions in all the three rotational degrees of freedom φ_X, φ_Y, φ_Z in connection with the elastic coupling of the mobile assemblies upon articulated solid joints are made possible by the redundant drives in the Y- and Z-axes. The machine frame of the test bed is designed in lightweight construction and consists of three steel angle sections. A light weight framework carries the milling spindle and enables force and torque measurements at the TCP in all directions by means of the load cells mounted into the rods (Fig. 1.4).

Test bed control was implemented by the PC based system TwinCAT 3.1 by Beckhoff. EtherCAT was used as the fieldbus. Path generation and data logging are performed by a higher-order user interface (BOF) developed at the IWM. User interface and TwinCAT communicate via ADS interface (Automation Device Specification). In this communication, ADS makes available ports for various areas in the TwinCAT system. Thus, the desired path is specified and, conversely, data are logged via an ADS port to the TwinCAT's PLC. For determination of the temperature field, 36 Pt100 temperature sensors were applied alone to the X-, Y- and Z-slides. All measured data were captured with EtherCAT modules and can be logged centrally in the machine control (BOF).

The MiniHex is used as an additional demonstrator of smaller size. It is fitted with six struts of variable length. The struts are driven by ball screws, each with a synchronous servo-motor. The positions were measured indirectly upon the servo-motors' rotary transducers. Cardan joints of simple structure are used as articulated joints between the base frame and the strut axes or the strut axes and the platform.

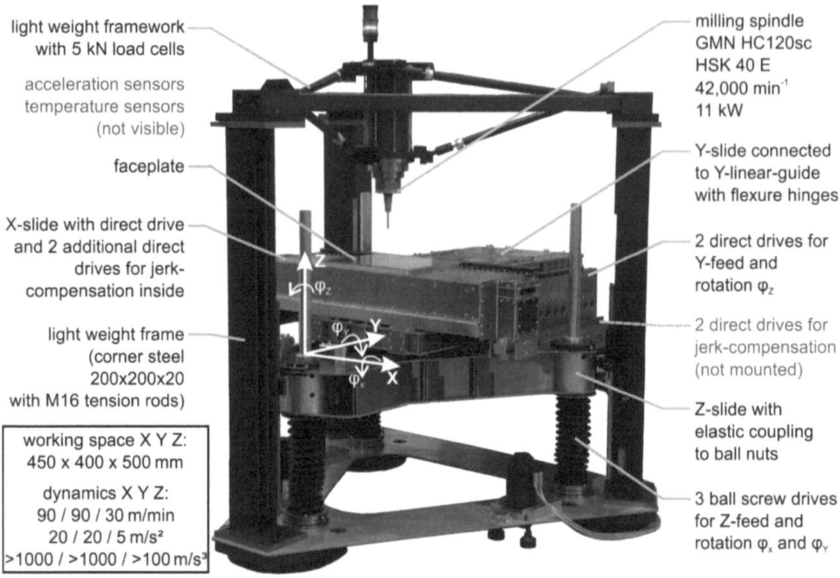

Fig. 1.4 Test bed at the Dresden location

The rotational degree of freedom, which is required for the kinematic mechanism of the Hexapod, is implemented by the torsion of the ball nut against the strut axes' screw shaft. Control of the MiniHex and capturing of the measured temperatures is executed in the same way as for the test bed with the PC based controlling system TwinCAT 3.1.

In Chemnitz, there is an *experimental machine*, whose structure conforms to a commercially available machining centre, but which by contrast enables manifold partial modification options through the exchange of assemblies and integration of components. As a result, this machine is extremely well suited for testing and the investigation of the compensation solutions.

In Aachen, explorations are performed under real manufacturing conditions similar to practice on two *production machines* representing the current industrial standard—a milling centre (DBF 630) and a grinding machine (Microcut BD 10).

The following chapters elucidate in detail the strategy and the outcomes of the 19 subprojects of the CRC/TR 96, whose exploration began in July 2011 and has continued until now.

All authors would like to thank the German Research Foundation (DFG) for financial support within the CRC/TR 96.

References

Brecher C, Fey M, Bakarinow K, Neus S (2014) Kugelgewindctrieb-Drchzahl bccinflusst Maschinengenauigkeit. MM Maschinenmarkt 27:60–63

Kinkel S (2005) Anforderungen an die Fertigungstechnik von morgen. Mitteilungen aus der Produktionsinnovationserhebung. PI-Mitteilungen 37

Kinkel S, Lay G (2006) Technologietrends in der Produktion. Praxis der Anlagenmodernisierung in der deutschen Metall- und Elektroindustrie. Mitteilungen aus der Produktionsinnovationserhebung 39

Chapter 2
Model-Based Representation of Thermo-energetic Effects in Cutting Tools and Part Clamping Devices

Michael Bräunig, Ulrich Semmler, Gerhard Schmidt, Volker Wittstock and Matthias Putz

Abstract In the representation of the machine tool's manufacturing accuracy, involving thermo-elastic deformations in the tool and its chuck in the considerations is unavoidable, since a portion of the process heat is transmitted into the tool. Capture and estimation of the resultant thermo-elastic deformations is of significant importance for their compensation and correction. This paper presents the investigation and evaluation of thermo-elastic displacements in the tool-chuck system resulting from the heat flows from machining. The temperature and displacement measurements were conducted on a test bed with measuring equipment. In the first series of experiments, the transient heat conduction in the stationary tool and the chucking system, decoupled from the motor spindle, was captured. The chucks were compared in terms of their working principle. The identification of the thermal boundary and contact conditions was assisted by Finite Element models. The paper describes the experimental setups, the methods of measurement and the numerical models. An additional section discusses the determination of realistic heat sources generated in the machining process. Since measurement of these heat flows is either impossible or expensive and time-consuming, they are determined by numerical simulation of the machining process itself. This way, it can be determined to what extent the heat flows generated in the experiments by means of heating cartridge or induction correspond to those generated by the actual machining heat sources.

M. Bräunig (✉) · V. Wittstock · M. Putz
Faculty of Mechanical Engineering, Chemnitz University of Technology, Chemnitz, Germany
e-mail: michael.braeunig@mb.tu-chemnitz.de

V. Wittstock
e-mail: volker.wittstock@mb.tu-chemnitz.de

M. Putz
e-mail: matthias.putz@mb.tu-chemnitz.de

U. Semmler · G. Schmidt
Fraunhofer Institute for Machine Tools and Forming Technology, Chemnitz, Germany
e-mail: ulrich.semmler@iwu.fraunhofer.de

G. Schmidt
e-mail: gerhard.schmidt@iwu.fraunhofer.de

K. Großmann (ed.), *Thermo-energetic Design of Machine Tools*,
Lecture Notes in Production Engineering, DOI 10.1007/978-3-319-12625-8_2

2.1 Introduction

2.1.1 Motivation and Problem Definition

The machining heat generated immediately in the tool-workpiece contact area is a major heat source in the overall system of the machine tool. Heat dissipates into the tool chuck and from there into the motor spindle, as well as through the part and the part clamping system into the machine table. The temperature fields cause non-negligible thermo-elastic deformations in the systems tool chuck—spindle, as well as the part clamping system—machine table that work against the requirement for manufacturing accuracy. The trend towards high performance cutting (HPC) and dry metal cutting or minimal quantity lubrication increases the thermal stresses and thus thermally-induced workpiece errors. The conflict among the criteria of accuracy, productivity and energy consumption is complicated by small manufacturing batches, changing manufacturing orders and tool changes, as well as shut downs of the base load power to conserve energy. The result would be a thermally unstable machine tool structure with low manufacturing accuracy.

2.1.2 Aim

The aim of the subproject A01 on "Deformation of tool and clamping device" consists of modelling the tool and its clamping system (chuck) under the assumption of real process parameters, such as process heat and the self- and externally-excited bypassing flows as the input variables, the boundary and environmental conditions of the workspace, and the heat transfer conditions at the relevant interfaces of the tool and its clamping system.

The exploration of the thermal manufacturing influences for complex metal removal processes is the subject of the DFG main programme 1480. The general aim is modelling and simulation of, as well as compensation on, the component or workpiece through experimental and simulation assisted investigations (Biermann and Iovkov 2013). Subproject A01 limits this overall focus to tool-oriented studies.

Experimental and model-assisted investigations are conducted to identify the relevant parameters, such as temperature, dislocations and heat flows. The challenge in the experiment is to realise the process heat, whereas, in simulation, assumptions regarding the convective heat transfer conditions have to be agreed. The two methods of investigation—experiment and simulation—supplement one another and provide a verification of the parameters mentioned above.

The identification of the temperature fields makes it possible to abstract compensation measures in order to obtain stable thermo-elastic characteristics in the future.

Fig. 2.1 Test bed equipped with measuring equipment

2.2 Approach

2.2.1 Test Bed Description

A test bed equipped with measuring instruments was designed and built (Fig. 2.1) to investigate the thermal behaviour of the cutting tools and the clamping systems—that is, the tool chuck and part clamping device—experimentally. The placement of the setup in an air-conditioned room allows for long-term-experiments with temperature monitoring. The test bed consists of machine bed, motor spindle to be mounted both in horizontal and vertical positions, control, cooling lubrication system and measuring equipment. In addition to the influence of the self-circulating and forced moving air on the thermal behaviour, the influence of the cutting fluid can be shown.

2.2.2 Subject of Experimental Investigations

A 20 mm diameter end mill made entirely of carbide was used as the tool. The milling cutter can be integrated into the test bed, has cooling channels inside, and allows for transferability of the results to other tools and machining processes, such as drilling. For tool clamping, three tool chuck principles were investigated in detail: hydraulic chuck, shrink fit chuck and chuck with a lateral clamping surface

Fig. 2.2 Measurement setup for the rotating tool

(Weldon type chuck) each with SK40 spindle interface. Up to 16 measuring points for temperature and 4 measuring points for deformation were attached to the tool and the chuck. Temperature measurement was predominantly executed by resistance thermometers designed using thin-film technology as Pt100 or Pt1000 sensors. The displacements were traced by inductive sensors, eddy current sensors and laser sensors. For tools in standstill position, the measurements were conducted at the SK40 interface, whereby the tool was clamped in the motor spindle and thermally insulated at the SK40 interface. Rotating tools were measured under the influence of motor spindle tempering. For rotating tools, the temperature measurement signals were transmitted to the measurement amplifiers by means of slip ring assemblies. Contactless eddy current transducers and laser sensors were employed exclusively. This decision made it necessary to have available planar measurement surfaces on the tool. To fulfil this condition, initially a dummy end mill was employed in the experiments. The transfer to the real tool's geometry was conducted using the simulation assisted investigations. The measurement setups for the various operation modes are described in detail in (Drossel et al. 2013). Figure 2.2 illustrates the measurement installation applied to the rotating tool.

For the investigations of the tool in standstill position, the tool was heated using a ceramic immersion heater. Effective heat transfer was provided by thermally conductive grease, so that heat immediately dissipated into the tool. In contrast, for rotating tools, the principle of contactless inductive heating was employed.

2.2.3 Simulation Assisted Investigations

In the measurement setup, the real amount of the heat flow introduced and the parameters for heat transfer to the environment or between tool and chuck, as well as between the chuck and the motor spindle were unknown. For their identification, FE models describing the tools employed (real tool and dummy tool), as well as of all chucks, were developed, and the parameters were adjusted by making use of the characteristic temperature and deformation parameters measured. Figures 2.3 and 2.4 illustrate the respective FE models, as well as the thermal boundary and contact conditions underlying the models.

The thermo-elastic and thermo-physical material parameters of the K40 tungsten carbide (tool material) and 1.2342 (X38CrMoV5-1) (chuck material) are taken from the material and manufacturer's information databases.

Thanks to these validated models, arbitrary thermal loads—varying in amount and timing—can be simulated without expensive and time-consuming experiments. This strategy targets simplified models for simulation in real time, capable of delivering the current correction values during the manufacturing processes.

The measurement and simulation results are introduced and discussed in the following section.

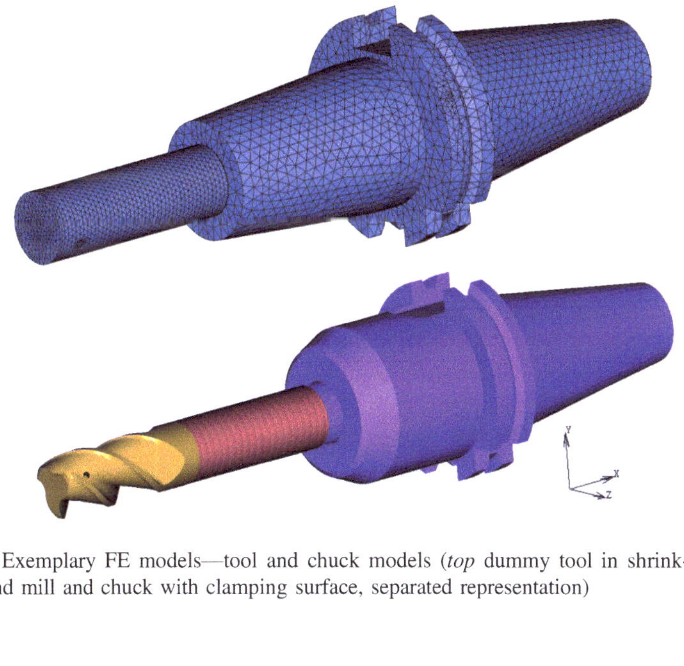

Fig. 2.3 Exemplary FE models—tool and chuck models (*top* dummy tool in shrink-fit chuck, *bottom* end mill and chuck with clamping surface, separated representation)

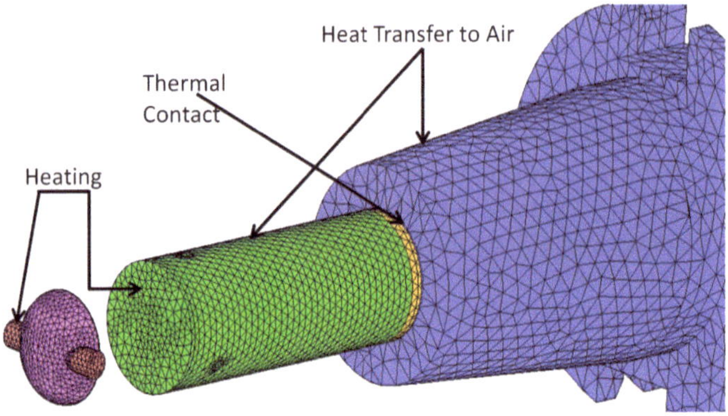

Fig. 2.4 FE model of dummy tool in shrink-fit chuck, with thermal boundary conditions and heat transfer between tool and chuck (from tool to chuck)

2.3 Results

2.3.1 Temperature Fields and Thermo-elastic Elongation in the Chuck

Initially, the clamping principles were compared in a preliminary test series. These tests had to verify the measurement methods and reliable repeatability of measurements in each test setup, as well as a relative comparison of the clamping principles, taking into account relevant influential factors in the real chip forming process. Figure 2.5 outlines the values provided for the stationary state of the three chuck types used.

Analyzing the shrink-fit chuck, a homogeneous drop in temperature along the rotational axis can be observed. For the process parameters mentioned, tool tip displacement amounts to approximately 60 μm. The hydraulic chuck suggests that the hydraulic clamping principle isolates the tool from the chuck, because the temperatures inside the tool are clearly higher. Significantly lower displacements can be detected on the chuck with a lateral clamping surface. Geometry has an important influence on the thermo-elastic elongation. The chuck is smaller in size and cannot be directly compared with the other clamping principles. The tool is arrested by means of a setscrew, and an air gap emerges as a result. The air gap has an insulating influence on the tool, and, consequently, less heat is transferred to the chuck. The temperatures in the chuck are less versus shrink-fit- and versus hydraulic chucks.

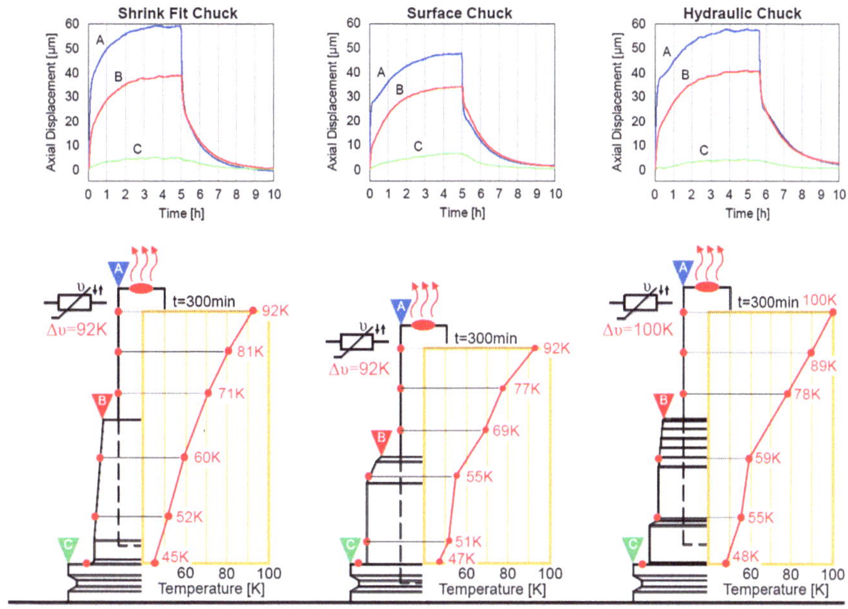

Fig. 2.5 Comparison of the displacement values measured and the stationary temperature profiles for shrink-fit chucks, chucks with damping surface and hydraulic chucks

2.3.2 Simulation Results and Adjustment with Experimental Investigations

The thermal boundary and contact conditions for the test series with the standstill tool were exemplarily adjusted for the shrink-fit chuck. The closest correspondence with the measurements was achieved for a heat transfer coefficient $\alpha_{Luft} = 8\,\mathrm{W/m^2K}$ to the environment and a heat flow of 8.1 W into the tool's tip. Figures 2.6 and 2.7 show the measured and simulated temperature values for these parameters, as well as the axial displacements at the tool tip and on the chuck's face. The temperature measuring points TMP1–TMP5 correspond to the measuring points in Fig. 2.5—from tool tip to chuck.

In the stationary temperature range, the absolute deviations between the values measured and calculated amounted to a maximum of 4 K in the case of temperatures and 7 µm in the case of displacements.

The contact heat transfer coefficients in the tool-chuck contact cannot be traced directly. In the FE models, based on investigations of the subproject B02, a value of $\alpha_{therm} = 12{,}000\,\mathrm{W/m^2K}$ (see Chap. 11 and Ustinov et al. 2010) was used. However, these coefficients turned out not to play a significant role in the heat flow out of the tool, as can be seen for different α values by the temperature curve along the interface (Fig. 2.8).

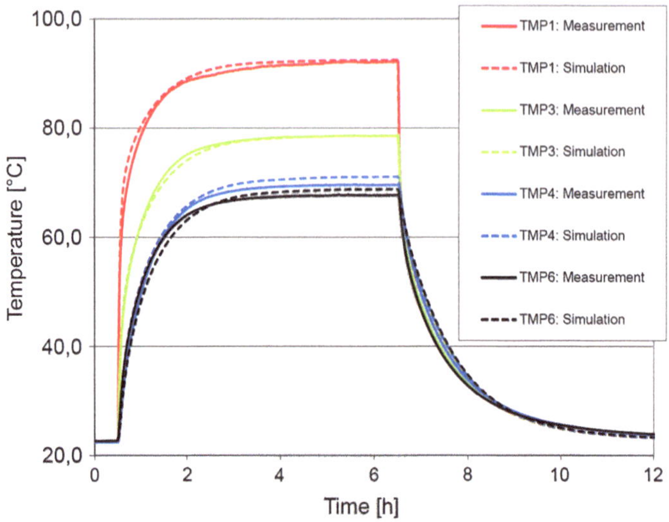

Fig. 2.6 Measured versus simulated temperatures for the shrink-fit chuck with clamped dummy tool

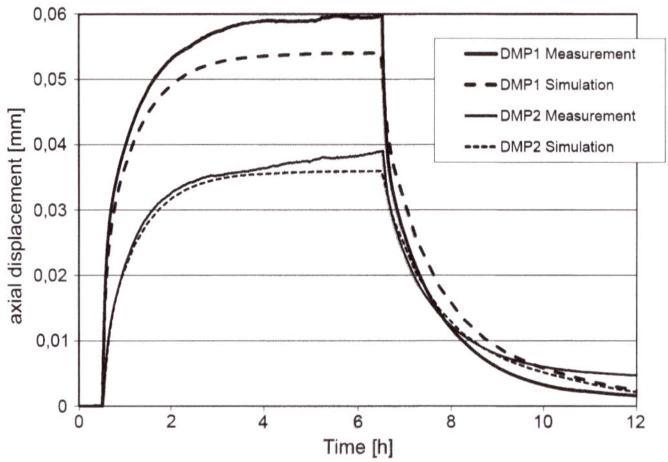

Fig. 2.7 Measured versus simulated displacements (*DMP1* face side of tool, *DMP2* face side of chuck)

The temperature jumps are shown for two cross sections, which correspond to the positions of the temperature measuring points TMP4 and TMP5 (upper curves 20 mm, lower curves 40 mm distance to face of the chuck).

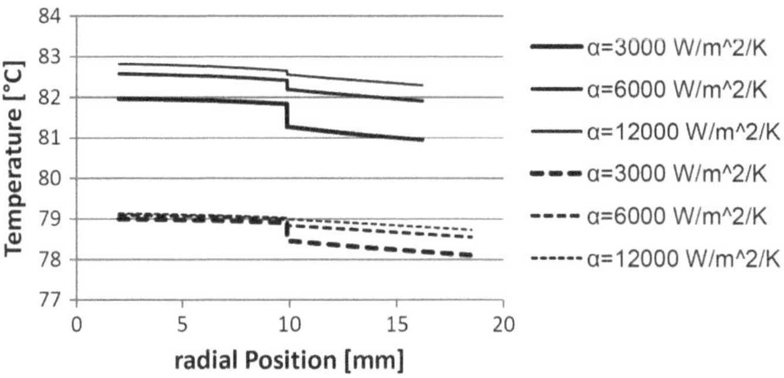

Fig. 2.8 Temperature jumps calculated in the tool-chuck contact for different heat transfer coefficients α

2.3.3 Determination of Process Heat Parameters

The amounts of heat generated in the machining process, their introduction in the form of heat flows dissipating into the tool, the workpiece and the chips (which, in turn, also introduce heat into the machine tool at the point at which they are deposited or removed) and the TCP displacements to be corrected or compensated are of crucial importance for thermo-energetic machine tool design. It has still to be verified how accurately the heat sources in the measurements, either by ceramic immersion heater or by induction heating, represent realistic sources, and, consequently, to what extent the TCP displacements measured or simulated are representative for the system of the chuck-tool-part-clamping device. Complex measurement equipment is needed to directly measure the heat content generated in the machining process, and measurements are only possible for simplified process conditions with tools and parts of adequate design (Davies et al. 2007).

For temperature distribution in the chip formation and friction contact zone, empirical and analytical models are provided by Komanduri et al. (2001). However, all of these models are based on simplified model assumptions and are thus inaccurate and only transferable to a limited extent to most machining processes. Consequently, the input variable from process heat was selected so that the parameter is located in the range of conventional process representation and can be captured by means of thin-film sensors. This means that the absolute temperatures at the measuring points, particularly those located immediately behind the chip formation zone, should be less than 120 °C.

The heat source affecting a real machining process results from the plastic deformation energy transformed during chip formation and the frictional heat generated by the contact of tool and workpiece or chip. For heat source determination, numerical simulation of the chip formation procedure was used. The methodology is described in detail by Drossel et al. (2014). Simulation results are shown for machining of steel AISI-1045 in free orthogonal cut with the model parameters following:

cutting speed $v_c = 10, 35, 70, 140, 280, 560\,\mathrm{m/min}$,
depth of cut $h = f_z = 30\,\mu\mathrm{m}$, cutting edge radius $r = 5\,\mu\mathrm{m}$,
rake angle $\alpha = 2^\circ$,
tool orthogonal clearance $\alpha = 7^\circ$.

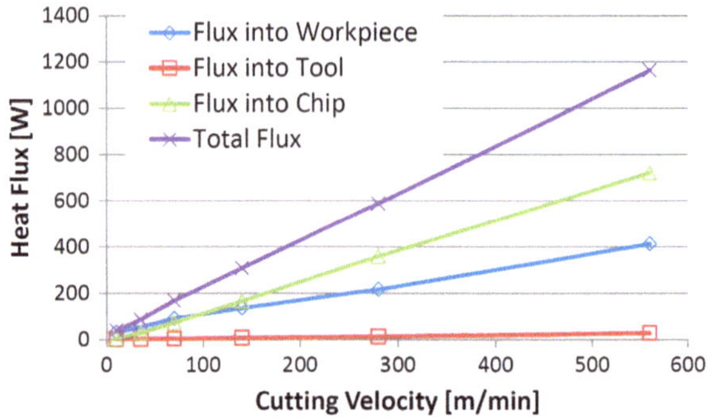

Fig. 2.9 Heat flows calculated for different cutting speed values

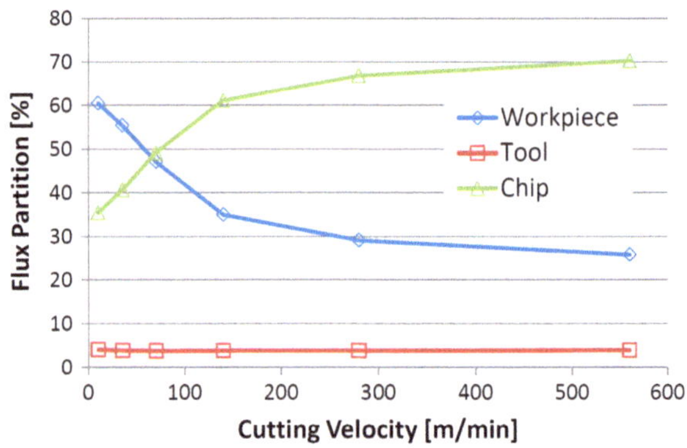

Fig. 2.10 Heat flow distribution in per cent, calculated for different cutting speed values

These parameters are referred to in Figs. 2.9 and 2.10.

Concerning the thermal balance of the machining process, heat generated in the machining process is increasing approximately proportionally with cutting speed (Fig. 2.9), whereby the fraction of the process heat that dissipates into the chips grows substantially (approximately 70 % for $v_c = 560\,\mathrm{m/min}$, Fig. 2.10), whereas the portion dissipating into the workpiece drops down to 26 %. For the consequences in the tool-chuck system investigated here, the tool fraction of process heat is important, which only slightly changes with increasing cutting speed and which is approximately 4 %.

For other machining processes, such as turning (non-free cutting), milling and drilling, the heat flow values P_{sim} calculated in simulation have to be scaled with factors or correction values according to Table 2.1:

Table 2.1 Correction values for calculation of heat dissipating from machining

Parameter	Meaning
$C_1 = \frac{A_{chip\_real}}{A_{chip\_sim}}$	Other chip cross section
$C_2 = z$	z cutting edges engaged
$C_3 \approx 1.5$	3D process (for instance external cylindrical turning) instead of simulated 2D process
$C_4 \approx 0.75$	Process with varying chip area (milling)
$C_5 = \frac{t_{cut}}{t_{total}}$	Process with interrupted cut

$$P_{real} = P_{sim} \cdot C_1 \cdot C_2 \cdot C_3 \cdot C_4 \cdot C_5 \tag{2.1}$$

For an HPC milling process using the end mill, as in the experiments (d = 20 mm, 3 cutting edges) with the cutting parameters n = 2,228 min^{-1} (that means v_c = 140 m/min), a_e = 5 mm, a_p = 2 mm, f_z = 0.05 mm (that means A_{chip_real} = 0.1 mm^2), the following correction values are obtained (A_{chip_sim} = 0.03 mm^2) $C_1 = 3.33$, $C_2 = 3$, $C_3 = 1.5$, $C_4 = 0.75$, $C_5 = 0.167$. Simulation of the 2D planing process results in a heat flow P_{sim} = 13.1 W, which results in a real heat flow of P_{real} = 24.6 W under the consideration of the correction values.

The thermal output of this heat source is approximately 300 % of the heat source of 8.1 W, which was used in the long-term experiments. In the real processes, these thermal impacts mostly appear over shorter periods of time; however, for instance when machining integral parts made of titanium, they can last longer. Taking this into account, the displacements of the experiments have realistic dimensions.

2.4 Classification of Outcomes in the SFB/TR 96

The temperature and deformation characteristics' models of the tool and the chuck —and in the future also of the workpiece and the clamping device, as well as the resultant heat flows into the motor spindle and the machine table—include coefficients of heat transfer between the parts clamped, resulting from explorations carried out in the subproject B02. Although, as has been demonstrated, these heat transitions are insignificant for the tool-chuck interface due to the high clamping forces, they will probably be relevant in the systems of part-clamping device—machine table and of chuck-motor spindle.

The force and temperature measurement results obtained on the test bed developed in subproject A02 (see Chap. 4) were used to adjust the simulation models for heat source calculation. Conversely, the simulation results are used to verify the analytical chip formation process model to be developed in subproject A02. The content and distribution of the machining heat is also important to know in this subproject. These results can and will be transferred to other, particularly

3-dimensional, cutting processes and processes with interrupted cut in the future (Klocke et al. 2011).

The numerical simulation method developed for the calculation of heat flows dissipating from metal cutting can be analogously transferred to machining processes with geometrically undefined cutting edge, like those investigated in the subproject A03. The proof of transferability has hitherto been given in the simulation of engagement of an individual grain, whereby the force and temperature measurements in subproject A03 are used for model adjustment. From the present state of expertise, it is impossible to determine the grinding process's heat source alone from the grain engagement temperatures measured alone, without the numerical simulation of the metal removal process.

2.5 Outlook

The measurements and simulations of the system tool chuck–spindle, without considering the influence of cutting fluids, have been completed. The simulation models considering the influence of air and cutting fluids on heat flow have been developed, but have yet to be explored regarding their effects on the thermo-energetic behaviour (Krettek 2014).

The evaluation of the identified parameters makes it possible to elaborate approaches for compensation, including, for instance, design optimisation through geometry and influence on heat flows through calculated measures.

Thermo-energetic investigations regarding the part clamping device will be performed in the future. As a precondition for these studies, a test bed was designed and installed, and the numerical models were developed.

The simulation methodology to balance the heat dissipated from metal cutting processes has to be extended to additional materials and manufacturing processes (turning as non-free cutting, milling, drilling). The correction factors C_1–C_5, which have been hitherto determined by means of exemplary models, have to be proven for these processes.

References

Biermann D, Iovkov I (2013) Modeling and simulation of heat input in deep-hole drilling with twist drills and MQL. In: 14th CIRP conference on modeling of machining operations, 13–14 June. Procedia CIRP, vol 8, pp 88–93

Davies MA, Ueda T, M'Saoubi R, Mullany B, Cooke AL (2007) On the measurement of temperature in material removal processes. CIRP Ann Manuf Technol 56(2):581–604

Drossel WG, Wittstock V, Bräunig M, Schmidt G (2013) Untersuchung der thermischen Werkzeugverformung. wt-Werkstatttechnik online 103(11/12):882–887

Drossel W-G, Schmidt G, Semmler U, Wittstock V, Bräunig M (2014) Thermal deformations of cutting tools: measurement and numerical simulation. Prod Eng Res Dev 8:543–550. doi:10.1007/s11740-014-0538-y

Klocke F, Bergs T, Busch M, Rohde L, Witty M, Cabral GF (2011) Integrated approach for a knowledge-based process layout for simultaneous 5-axis-milling of advanced materials. Adv Tribol 2:108–115

Komanduri R, Hou ZB (2001) Thermal modeling of the metal cutting process, part III: temperature rise distribution due to the combined effects of shear plane heat source and the tool-chip interface. Int J Mech Sci 43:89–107

Krettek L (2014) Simulation einphasiger Strömungsausbildung am rotierenden Fräswerkzeug. Master thesis, TU Chemnitz

Ustinov V, Kneer R, Al-Sibai F, Schulz SG, El-Magd E (2010) Influence of surface roughness on the contact heat transfer. In: The 14th international heat transfer conference IHTC14, Washington DC, USA, 7–13 Aug

Chapter 3
Model and Method for the Determination and Distribution of Converted Energies in Milling Processes

Matthias Brockmann, Fritz Klocke and Drazen Veselovac

Abstract This paper summarises the current state of project research focussing on the energy model for milling processes. The general methodology of heat flux determination is described in detail. An analytical temperature distribution model that has been engineered in a subproject is presented as an initial result. The potential theory provides complex solutions for the differential equation of thermal conduction that make available temperature field and heat flux field at the same time. The models were validated by means of the temperature fields measured. An online calibration method combining an infrared camera with a two-colour pyrometer was developed to ensure the quality of the captured data. Finally, the principal paradigm of determining heat fluxes from the temperature models is briefly introduced.

3.1 Introduction

Metal cutting using a geometrically defined cutting edge is a highly sophisticated physical process. The cutting edge penetrates the part material until a critical shear stress value has been overcome. The surface layer removed is then moved across the tool's rake surface. During the cutting process, most of the mechanical energy required is converted into thermal energy. Thus the metal cutting process is one of the main heat sources in a machine tool. The temperature distributions in the components involved in metal cutting—chip, part and tool—immediately affect the

M. Brockmann (✉) · F. Klocke · D. Veselovac
Laboratory for Machine Tools and Production Engineering, Faculty of Mechanical Engineering, RWTH Aachen University, Aachen, Germany
e-mail: M.Brockmann@wzl.rwth-aachen.de

F. Klocke
e-mail: F.Klocke@wzl.rwth-aachen.de

D. Veselovac
e-mail: D.Veselovac@wzl.rwth-aachen.de

K. Großmann (ed.), *Thermo-energetic Design of Machine Tools*,
Lecture Notes in Production Engineering, DOI 10.1007/978-3-319-12625-8_3

thermo-elastic deformation of tool and part holders, heat radiation into the machine tool's interior and the metalworking fluid's temperature. Consequently, the target of the subproject is the derivation of a heat flux distribution model based on parameters. In addition to the pure machining parameters, the parameters characterising different metal cutting processes and various combinations of part and tool materials, are also understood as parameters. The focus of the research is mainly on the milling process.

3.2 Approach

Fourier's law of thermal conduction is used to determine heat flux through thermal conduction:

$$q = -\lambda \frac{\partial T}{\partial n} \tag{3.1}$$

Heat flux q can thus be determined from thermal conductivity λ and the temperature gradient $\partial \mathrm{T}/\partial \mathrm{n}$ in each normal direction. Since it is impossible to measure the heat fluxes directly, the temperature fields appearing in the components of chip, tool and part during metal cutting are the key to the parameterised model that is sought. To derive the models for the temperature fields, in principal, it is possible to select empirical, analytical or simulation approaches. Empirical models are derived from measurements by means of regression without having to know the physical laws. In simulation and analytical models, these laws are taken into account in the solution to the thermal conduction differential equation. The thermal conduction differential equation can be presented in a generalised form as:

$$\rho c \frac{\partial T}{\partial t} = \left[\frac{\partial}{\partial x}(\lambda \frac{\partial T}{\partial x}) + \frac{\partial}{\partial y}(\lambda \frac{\partial T}{\partial y}) + \frac{\partial}{\partial z}(\lambda \frac{\partial T}{\partial z}) \right] + \Phi \tag{3.2}$$

For constant thermal conductivity λ, it may be simplified to:

$$\rho c \frac{\partial T}{\partial t} = \lambda \left(\frac{\partial^2 T}{\partial x^2} + \frac{\partial^2 T}{\partial y^2} + \frac{\partial^2 T}{\partial z^2} \right) + \Phi \tag{3.3}$$

where ρ is the density, c is the specific heat capacity and Φ a term to take into account heat sources in the control volume. As a prerequisite to derive temperature fields for metal cutting, mainly milling, the derived analytical solutions must satisfy the differential equation.

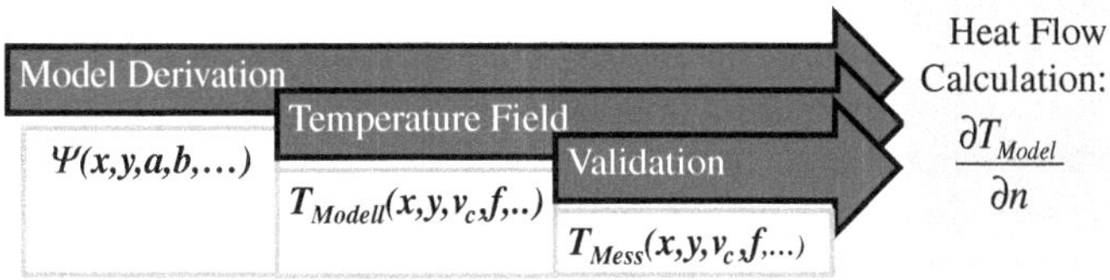

Fig. 3.1 Scientific approach for determination of heat flow

3.3 Results

Prior to the formulation of parameter-based temperature distribution and heat flux models in the metal cutting process, potential analytical solutions Ψ are derived, see Fig. 3.1. These solutions are initially the mathematical solutions to (3.3) and depend not only on the spatial coordinates x and y, but also the model parameters (in Fig. 3.1 named as a, b, …).

Having adequately identified these model parameters, it is possible to derive temperature distributions during metal cutting T_{Model}. The parameters in these distributions correspond to the cutting parameters (for instance, the cutting speed v_c or the feed f). The modelled temperatures are validated by comparing them with the captured values T_{Mess} using the temperature fields found experimentally. Only temperatures are measured rather than heat fluxes. For this reason, the heat flux field can only be validated indirectly. The heat fluxes are obtained from (3.1), differentiating the temperature field equation $\partial T_{Model}/\partial n$ in the correspondingly chosen normal direction.

3.3.1 *Derivation of Analytical Temperature Models in Metal Cutting*

Since sensor access to the measured objects is limited, models predicting the temperatures during metal cutting were required very early on. Most of the existing models are empirical ones, and frequently the so-called heat partition coefficient describing the percentage of heat that flows in the tool and the chip is used. Empirical models do not have a physical basis, and it is impossible to predict entire temperature fields. Due to these weaknesses, this type of model is inflexible and difficult to transfer to new cases of application. In contrast, analytical models are based on the solution of the thermal conductivity differential equation and are thus regarded in principle as universal approaches. The model by Komanduri and Hou (2000, 2001a, b) whose approach uses an inclined band heat source, is the last model in this context and includes former models in a special solution. All models that have been provided up to now are based on real solutions of the thermal conductivity differential equation.

In the modelling performed in subproject A02, a potential theory-based model was made available. The potential theory provides methods for a complex solution of the thermal conductivity differential equation. The advantages in comparison with the state of the art result from the simultaneous modelling of the temperature field and the heat fluxes, as well as the flexible modelling opportunities following the tool-kit principle. Using the methodology shown in Fig. 3.1, it is possible to derive a solution for the temperature field in metal cutting. The following complex function is obtained:

$$\begin{aligned} F(z) &= F_{uniform} + F_{vortex} + F_{corner} \\ &= Aze^{-i\alpha} + \frac{Bi}{2\pi}\ln(z + z_{sum}) + C \cdot z_{rot}^{k} \end{aligned} \tag{3.4}$$

The function is superpositioned from the partial solutions $F_{uniform}$, F_{vortex} and F_{corner}. Each function can be assigned a physical interpretation analogously to fluid mechanics, which need not be detailed here. For a detailed derivation, see Gierlings and Brockmann (2013). The imaginary part Ψ, which can be interpreted as a function for the temperature field according to the methodology in Fig. 3.1, is given as:

$$\begin{aligned} \Psi_{total}(x,y) = A\sin(\alpha) - \frac{B}{2\pi}\ln\left(\sqrt{x_{total}^2 + y_{total}^2}\right) \\ + C \cdot \left[\sqrt{(x\cos\theta - y\sin\theta)^2 + (x\sin\theta + y\cos\theta)^2}\right]^k \sin(k\theta) \end{aligned} \tag{3.5}$$

The parameters A, α, B, C, θ and k can be regarded as constants.

The terms x_{total} and y_{total} are introduced to take into account the heat sources from friction and shearing. Figure 3.2 presents a plot of the calculated temperature field as well as of an exemplary temperature curve along the contact zone. To validate the temperature fields measured, the temperature fields are compared with the measurements. The direct relationship between the solution parameters and the cutting parameters was not determined in the project activities demonstrated; the parameters were chosen so that the qualitative curve of the temperature fields conforms.

3.3.2 Measurement of Temperature Fields in Metal Cutting

Different commonly available temperature sensor systems were engineered for a metal cutting application to capture temperatures in the metal cutting process. First, indirect measurement techniques and techniques with time-resolution can be distinguished. Indirect measurements, such as the use of temperature-sensitive colours or of powders with a constant melting point, as well as the estimation of metallurgical microsections, are not feasible as temperature measurement techniques applied to science and industry.

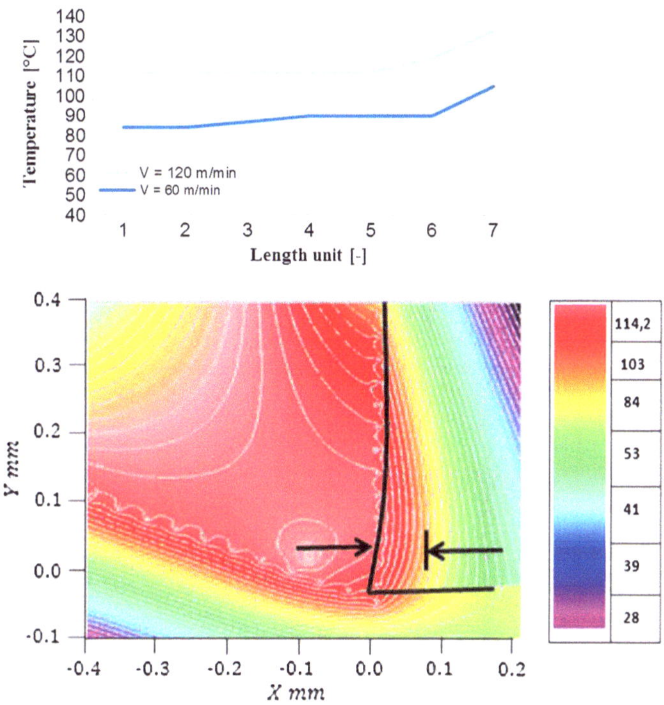

Fig. 3.2 Model plot and temperature along the contact zone

The more suitable time-resolved measurement methods, in turn, can be subdivided into thermo-electrical techniques and radiation measurement procedures. Resistance thermometers and thermocouples belong to the first type. These measurement methods were not employed in the subproject, since they are inappropriate for exact measurement of temperature fields in short measurement times.

The thermographic and pyrometric principles categorized as radiation measurement, offer the strength of contactless measurement, which can be distinctly advantageous in metal cutting investigations. Thermographic cameras are capable of recording a complete temperature image, with which it is possible to draw conclusions about a complete temperature field in a single measurement. One disadvantage of the camera technologies currently available, however, is that emissivity of metals, which, in turn, depends on wave length and temperature, is unknown. A real remedy is provided by two-colour pyrometry, based on—in contrast to thermography—the measurement of radiation at two discrete wavelengths. A suitable wavelength selection can allow for a change in emissivity to be neglected to a certain degree and thus guarantee absolute temperature measurement. However, a weakness of this technique is that the measurement is only conducted at one measuring point.

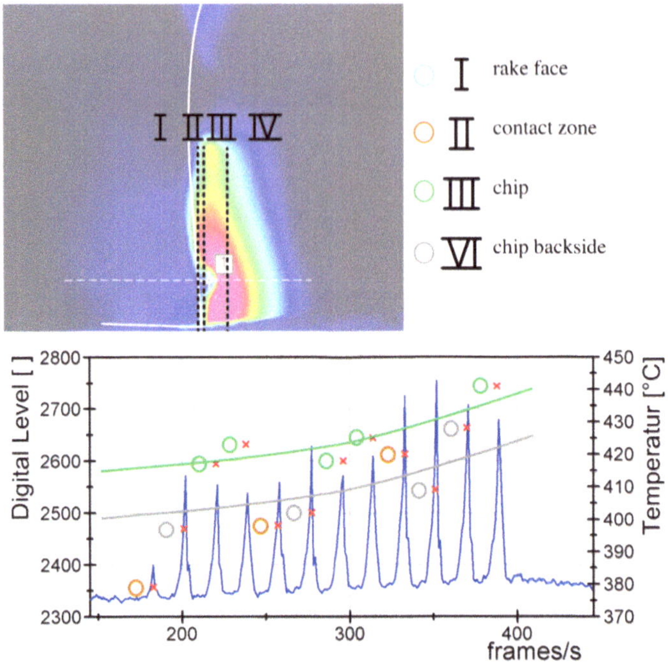

Fig. 3.3 Methodology for online emissivity calibration

To combine the strengths of the two radiation measurement techniques, an online emissivity calibration method was engineered in the subproject: a spot on the test object is recorded by both sensor systems at the same time. Thus it is possible to compare the digital level of the infrared (IR) camera with the temperature recorded by means of the two-colour pyrometer and use the outcomes for emissivity determination; compare with the diagram in Fig. 3.3.

Since the image rate is limited and dependent on the integration time, it is necessary to specify the location of the common measurement spot in the metal cutting zone. This is especially valid for the zones shown in the IR camera image—rake face, contact zone, chip and chip underside; see upper part of Fig. 3.3. By adjusting the IR images calibrated in the abovementioned manner, model validation by means of accurate measured data can be guaranteed.

3.3.3 Equation for the Heat Flux

To derive the heat flux model based on parameters from the temperature models, it is first necessary to derive the direction of the heat flux. According to (3.1), heat flux is to be determined in the corresponding normal direction. In other words, the

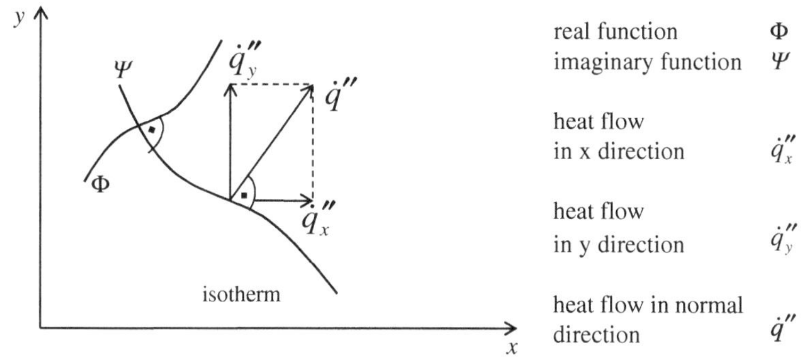

Fig. 3.4 Heat source field and direction

field is perpendicular to the isothermals determined by (3.5). Since the equation was derived from a potential function $F(z) = \Phi + i\Psi$, the real and imaginary parts of the function are in a mutually vertical position (except in singular points in which the Cauchy-Riemann conditions are not fulfilled). Consequently, the heat flux resulting from the real part Φ of (3.5) is given as:

$$\begin{aligned}\Phi = \mathrm{A}\cos(\alpha) + \frac{\mathrm{B}}{2\pi}\tan^{-1}\left(\frac{y_{total}}{x_{total}}\right) \\ + C[\sqrt{(x\cos(\theta) - y\sin(\theta)^2 + (x\sin(\theta) + y\sin(\theta)^2}]^k\cos(k\theta)\end{aligned} \tag{3.6}$$

The meaning of the parameters is identical with those in (3.5). The strength of the heat flux results from the gradient in the direction of this field. Figure 3.4 elucidates the geometric relationships between the isothermal and the heat flux.

The heat fluxes in x and y directions can be derived from the corresponding derivatives of the Ψ function. To calculate heat flux into the components, heat flux is multiplied by the area considered.

3.4 Classification of Outcomes CRC/TR 96

Parameter-based modelling of the temperature field induced by the metal cutting process is a key parameter for compensation and correction solutions.

The tool holder and part clamping components are immediately affected by the heat flux out of the metal cutting process. Heat is mainly transmitted via contact grooves between tool and tool holder as well as between part and part clamping. The machine interior is also affected directly by heat radiation, which is, in turn, dependent on the temperature of the component's chip, tool and part. The cooling lubricant that is available in most metal-cutting processes is also heated by convection and thus changes the temperature in the hydraulic circuit. As a whole, complex process representation is created, showing the tool and the part interacting with the cutting process as the heat source to be characterized by parameters (that is, to be changed by cutting parameters, such as cutting speed and feed).

This representation can be employed for process optimisation, on the one hand. On the other hand, the models can be used for specific relationships between energy input and accuracy curves as a function of productivity. In this way, they provide technological support for the overall system of the CRC/TR 96.

3.5 Outlook

The model to predict the temperature fields in the components of tool, part and chip engineered in the subproject A02 needs to be further developed in the ongoing subproject activities. In a first step, the relationship to the cutting parameters is generated explicitly in order to assess the models according to the productivity axis. In a second step, the model is expanded to more general chip geometries and cases of application. This is guaranteed by a method for the application of elementary solutions. Analogously to fluid mechanics, this paradigm is known as panel theory. Initially, the model can only be used for stationary cases in concrete applications. The intention is to extend the approach to unsteady cases by means of the theory of minor disturbances.

References

Gierlings S, Brockmann M (2013) Analytical modelling of temperature distribution using potential theory by reference to broaching of nickel-based alloys. Adv Mater Res 769:139–146

Komanduri R, Hou ZB (2000) Thermal modeling of the metal cutting process. Part I: Temperature rise distribution due to shear plane heat source. Int J Mech Sci 42:1715–1752

Komanduri R, Hou ZB (2001a) Thermal modeling of the metal cutting process. Part II: Temperature rise distribution due to frictional heat source at the tool–chip interface. Int J Mech Sci 43:57–88

Komanduri R, Hou ZB (2001b) Thermal modeling of the metal cutting process. Part III: Temperature rise distribution due to the combined effects of shear plane heat source and the tool–chip interface frictional heat source. Int J Mech Sci 43:89–107

Chapter 4
Energy Model for Grinding Processes

Matthias Rasim, Fritz Klocke and Patrick Mattfeld

Abstract A major percentage of the mechanical energy introduced during the grinding process is converted into heat. For this reason, the grinding process has a great influence on the thermo-elastic behaviour of the complete machine tool. This paper introduces an approach to model the energy conversion in the grinding process, taking into account grinding wheel topography. An analytical-empirical energy model for grain engagement is abstracted from the phases of chip formation when a single grain is engaged. Chip formation on a single grain was analysed using two newly developed test beds. In this approach, it was possible for the first time to identify the chip formation phases and to demonstrate their significant influence on the energy converted during grain engagement. Moreover, it was possible to derive initial characteristic parameters to model the influence of the grain shape. The Finite-Element simulation of single grain engagement makes it possible to vary the process variables in further experiments, in particular the grain shape properties, independent of one another and in a controlled manner, and to study their influence. It is the goal of ongoing studies to generalise the energy model for single grain engagement as a model for multi-grain engagement and thus the entire grinding process by means of a geometric-kinematic grain engagement model.

M. Rasim (✉) · F. Klocke · P. Mattfeld
Laboratory of Machine Tools and Production Engineering, Faculty of Mechanical Engineering, RWTH Aachen University, Aachen, Germany
e-mail: M.Rasim@wzl.rwth-aachen.de

F. Klocke
e-mail: F.Klocke@wzl.rwth-aachen.de

P. Mattfeld
e-mail: P.Mattfeld@wzl.rwth-aachen.de

K. Großmann (ed.), *Thermo-energetic Design of Machine Tools*,
Lecture Notes in Production Engineering, DOI 10.1007/978-3-319-12625-8_4

4.1 Introduction

In addition to the drives, the metal cutting process is a major heat source in the machine tool. In particular in the grinding process, due to the many small-sized cutting edges with negative rake angles, friction appears between the workpiece, tool and chips, which, in turn, results in heat. In addition to this, considerable heat is generated by the elastic and plastic deformation during chip formation (Brinksmeier 1982). The emerging heat dissipates into the components involved, that is, tool, part, coolant and chips. The amount and distribution of the heat fluxes vary as a function of process guidance and cooling strategy, as well as of the design of the components (Brinksmeier et al. 2006). The grinding process as a whole is complex due to the high number of irregularly shaped, statistically distributed grinding grains, which interact simultaneously with the part material. Consequently, at present, heat generation in the grinding process has been inadequately explored, and it is impossible to predict the amount of heat transferred to the machine by the process. However, this prediction is necessary in order to completely model the thermo-elastic behaviour of machine tools and to abstract solutions for workpiece errors arising from it.

For this reason, the overall target of the subproject A03 (see Chap. 3) is to develop a process model based on parameters that is able to represent the heat flux distribution in grinding. In the first step, the target is to calculate the relevant heat sources in the grinding process. In subsequent steps, the thermal conduction mechanisms for stationary and, later on, for unsteady conditions have to be identified. Finally, a model is shaped that maps heat generation in the grinding process and the distribution of the heat fluxes onto the components involved in the grinding process—grinding wheel, workpiece, coolant and chips.

4.2 Approach

4.2.1 Methodology to Model the Heat Sources in the Grinding Process

The major part of the mechanical energy introduced during the grinding process is converted into heat. Previous approaches to model heat generation in the grinding process consider the complete grinding wheel as a global input parameter and regard the grinding wheel in interaction with the workpiece in terms of its enveloping body. As a consequence, the models are only valid for a specific grinding wheel at a constant dressing condition. For generally valid modelling of heat generation, it is necessary to take into account the topography of the grinding wheel interacting with the workpiece. Grinding wheel topography can be regarded as a vast number of single grains interacting individually with the workpiece. Each chip formation that appears at a single grain represents a heat source (see Fig. 4.1).

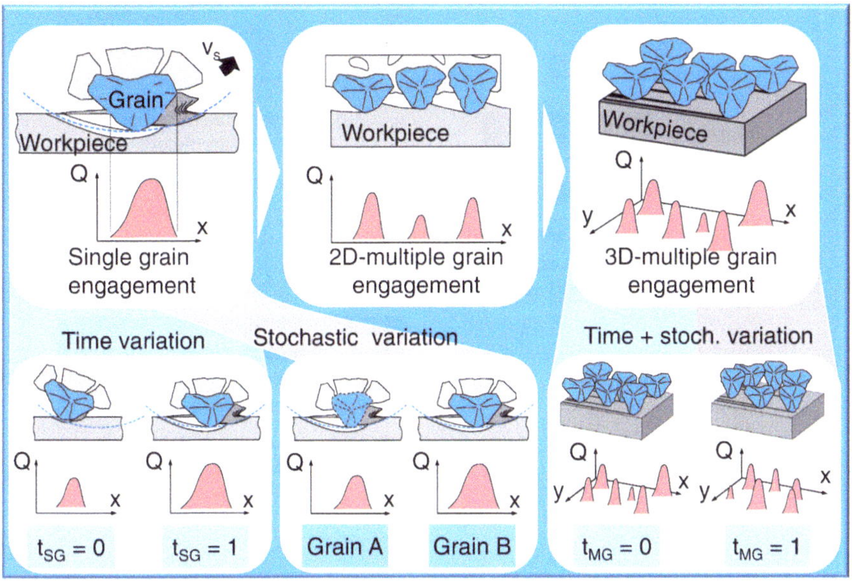

Fig. 4.1 Complexity of heat generation in the grinding process

The number of engaging grains, as well as their penetration depth, depends on the process parameters and the grain disposition on the grinding wheel. Correspondingly, when modelling heat generation in the grinding process, the approach that is followed is based on the chip formation procedures for the single grain. First, heat generation during the engagement of a single grain and the chip formation that results from this engagement are mapped as functions of the process parameters. To do this, experimental, as well as numerical, scratch tests with single grains based on the Finite Element method (FEM) are used. Based on these outcomes, the resultant heat source during the engagement of the grains in the overall contact zone will be modelled. Modelling will be conducted on the basis of a geometric-kinematic grain engagement model using scratch tests with multiple grains and grinding experiments (see Fig. 4.2).

4.2.2 Energy Model for Single Grain Engagement

In contrast to metal cutting processes with geometrically defined cutting edges, chip formation in grinding cannot be directly represented by the geometric-kinematic engagement conditions, since the grains, which are geometrically undefined, are characterised by complex engagement conditions. Model assumptions were developed that are able to map the chip formation process in grinding (Brinksmeier et al. 2006; Denkena et al. 2012). These models consider the path by which the abrasive

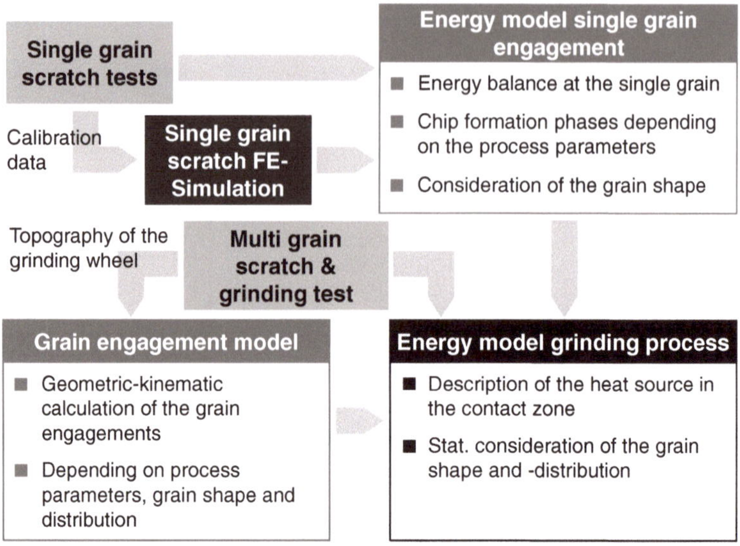

Fig. 4.2 Methodology of the subproject

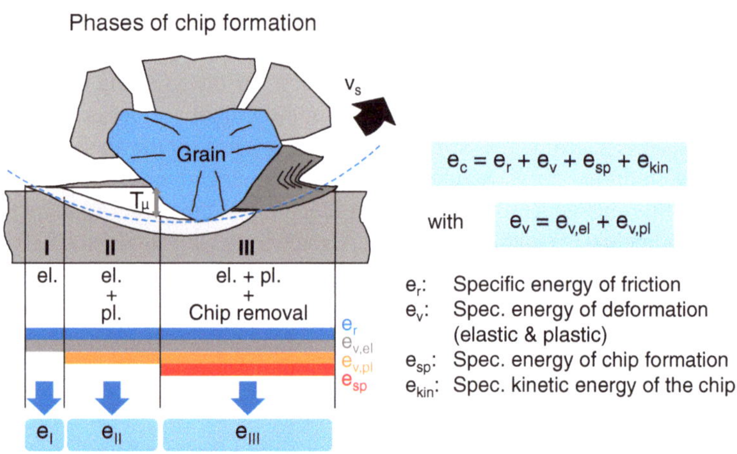

Fig. 4.3 Energy conversion during chip formation on a single grain (according to Klocke, Malkin)

grain penetrates the workpiece. Theoretical consideration of the grain engagement performed this way distinguishes zones of elastic deformation (I) and elastic-plastic deformation (II), as well as elastic-plastic deformation in combination with chip removal (III), see Fig. 4.3, (Marinescu 2004; Malkin and Guo 2008; Klocke and König 2005). Actual chip removal begins at the grain cutting depth T_μ, which, in turn, depends on the material to be cut, the grinding parameters used, and the friction conditions between abrasive grain and workpiece (Tönshoff et al. 1992).

Energy conversion in the grinding process can be classified according to the processes that occur during chip formation. Thus, the specific grinding energy e_c is composed of the specific energy of friction e_r, the specific energy of material deformation e_v, the specific energy of chip formation e_{sp}, the shearing energy, and the specific kinetic energy of the chip e_{kin} (Malkin and Guò 2008). The contribution of the chips' kinetic energy was shown to be less than one per cent and is thus negligible (Zeppenfeld 2005). The remaining fractions of the specific energy can be assigned to the individual chip formation phases (see Fig. 4.3). Friction, as well as elastic material deformation, occurs in all phases. The material is also subject to plastic deformation with the beginning of the second phase. Having achieved the grain cutting depth T_μ the workpiece material is cut off, and a chip forms. Consequently, it is possible to subdivide the specific grinding energy e_c into phase-specific part energies (e_I, e_{II}, e_{III}). In view of this division, according to (4.1), an analytic-empirical energy model for single grain engagement can be concluded.

$$e_c = e_I(a_I, b_I, \ldots) + e_{II}(a_{II}, b_{II}, \ldots) + e_{III}(a_{III}, b_{III}, \ldots) \tag{4.1}$$

The phase-specific partial energies, in turn, depend on different parameters (a_I, b_I, ...) that are defined by the materials and the process parameters, such as grinding wheel circumferential speed v_s, chip thickness h_{cu}, and the grain geometry. These influencing factors have to be determined by empirical investigations to build up the energy model.

4.3 Results

4.3.1 Investigations to Characterize Chip Formation

In the first step, chip formation and thus the development of the individual chip formation phases were analysed. The transitions between the individual chip formation phases depend not only on the process parameters, the part material and lubricant, but also mainly on grain geometry. A new test setup for single grain scratching was developed and implemented to identify the local resolution of the chip formation phases I–III during single grain engagement as a function of the kinematic process parameters and the grain shape (Rasim et al. 2013). In contrast to "conventional" single grain scratching, the grain to be scratched is here fixed on the machine table, and relative motion is carried out by the workpiece. Chip geometry was transformed in order to provide the same engagement conditions. To do this, the curve of cutting depth h_{cu} over the contact length was transformed into a convex contact length by superimposing the radius of a round solid (see Fig. 4.4). The contour calculated this way was ground into the workpiece by profile grinding, which was fixed afterwards on a rotating carrier disk. Since the scratching grain is in stationary position, it is possible for the first time to record the chip formation by means of a high-speed camera and high-frequency force measurement (see Fig. 4.4).

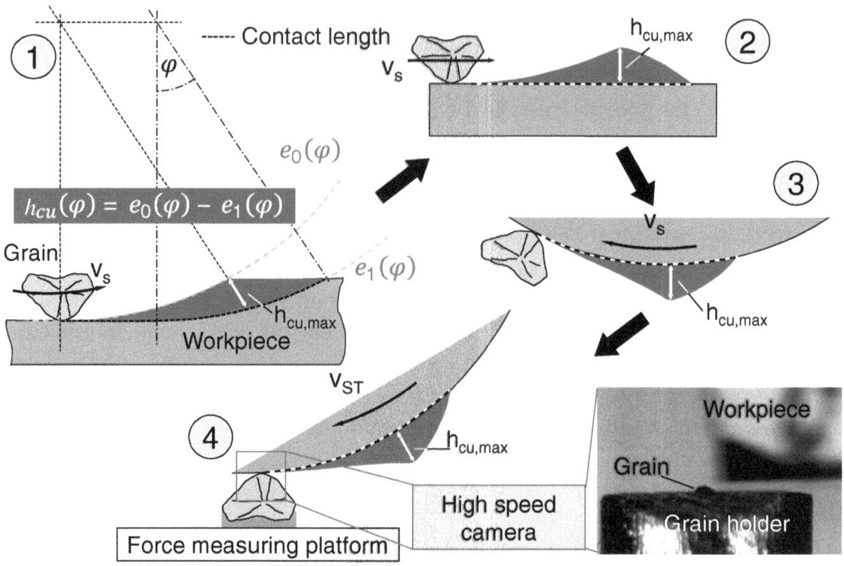

Fig. 4.4 Modified test setup for visual recording of the chip formation during single grain scratching (according to Rasim et al. 2013)

A great challenge in the camera images was the extremely short exposure time of less than $\Delta t_{expo} = 30$ µs resulting from the image frequency used, which was 33,333 fps (frames per second). Sufficient illumination of images could only be guaranteed by the use of high-performance light sources.

The chip formation phases can be visually identified by an analysis of the camera records. In the next step, the chip formation phases are correlated with the emerging forces and, in connection with the geometric engagement conditions, with the scratching energies. The high relevance of the chip formation phases for the description of the converted energy could already be shown in the results achieved so far. In Fig. 4.5, the curve of the instantaneous specific scratching energy s_e is shown. The curve is obtained from the quotient of the tangential force curve F_t and the curve of the groove section A_{cu} according to (4.2). The derivation can be taken from (Rasim et al. 2014).

$$s_e(t) = \frac{F_t(t)}{A_{cu}(t)} \tag{4.2}$$

As chip removal begins, a clear drop in the instantaneous specific scratching energy s_e can be observed (see Fig. 4.5), which is due to a change in the predominant mechanisms in the interaction of grain and part material. As expected, the energy necessary to remove a specific material volume is lower due to the transition from exclusively plastic and elastic material deformation into the shearing of workpiece material, which is conducted in combination with elastic and plastic deformations.

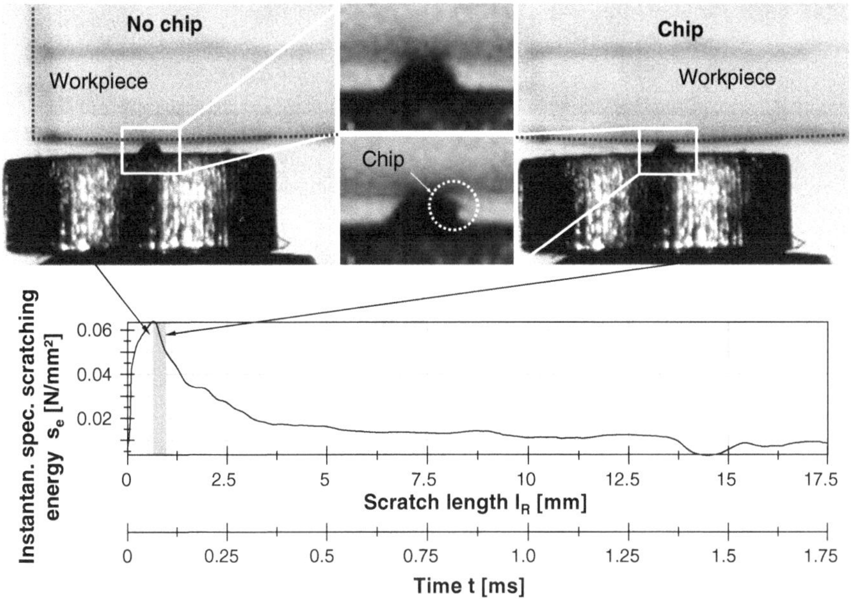

Fig. 4.5 Identification of beginning chip removal by means of high speed camera records and influence on instantaneous specific scratching energy

The database must now be expanded through additional investigations in order to determine influences of the grain shape and the process parameters. In particular, it is necessary to abstract characteristic parameters, which describe the characteristics of the grain shape that are significant for the formation of the chip formation phases, such as rake angle or section area, and can be used as parameters in (4.1).

4.3.2 *Analysis of Energy Conversion During Chip Formation*

In the second step, the energy conversion and thus the approximate heat generation during single grain engagement in each chip formation phase was analysed. In this analysis, the process forces that appear, as well as the geometric engagement conditions, are studied. To determine the influence of the process parameters and the grain shape on energy conversion, another test bed for the exploration of single grain engagement was developed. The target for this setup was to guarantee the high local and temporal resolution when recording the process parameters required by the engagement ratios. This setup makes it possible, on the one hand, to capture the process forces along the contact length of a couple of millimetres at a contact time of less than $t_k = 2$ ms, with good resolution. On the other hand, the temperatures that emerge can be measured with a foil workpiece-thermocouple at defined points (see Fig. 4.6). The measurement methodology employed made it possible to

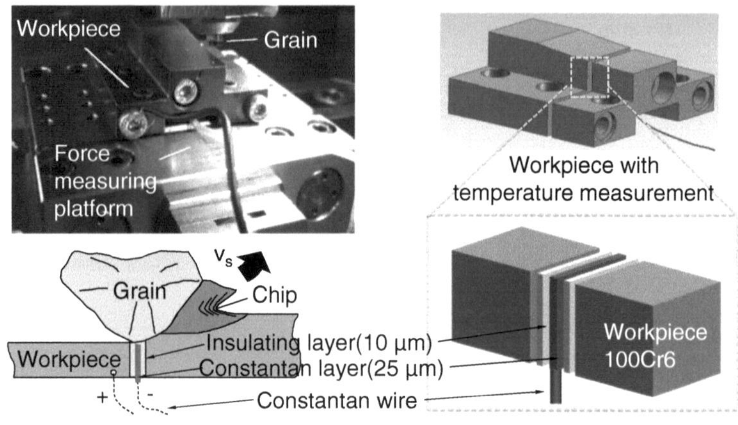

Fig. 4.6 Measurement method to record the temperatures during single grain scratching (according to Rasim et al. 2013)

successfully record the high temperature gradients of up to 10^7 K/s appearing during grain engagement (Rasim et al. 2013). The local temperatures determined this way were used for the qualitative assessment of the influencing factors on heat generation. Moreover, the temperature values can be applied as validation values for the Finite Element simulation illustrated in Fig. 4.8.

For the description of the grain shape influence, the above mentioned approach to abstract characteristic parameters that map the characteristics of the grain shape that are significant for energy conversion is followed. In Fig. 4.7, the temperatures are shown as a function of the groove section area A_{cu}. The temperatures increase continuously as a function of section area. A higher workpiece volume is deformed as a result of the greater groove section area. With increasing section area, the contact area between grain and part material also increases, so that more friction and, in turn, more heat is generated. In the investigations, the groove section area

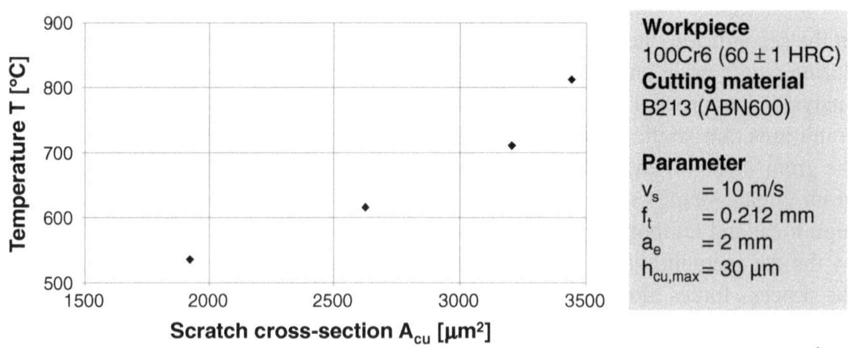

Fig. 4.7 Temperature depending on the scratching section during single grain scratching

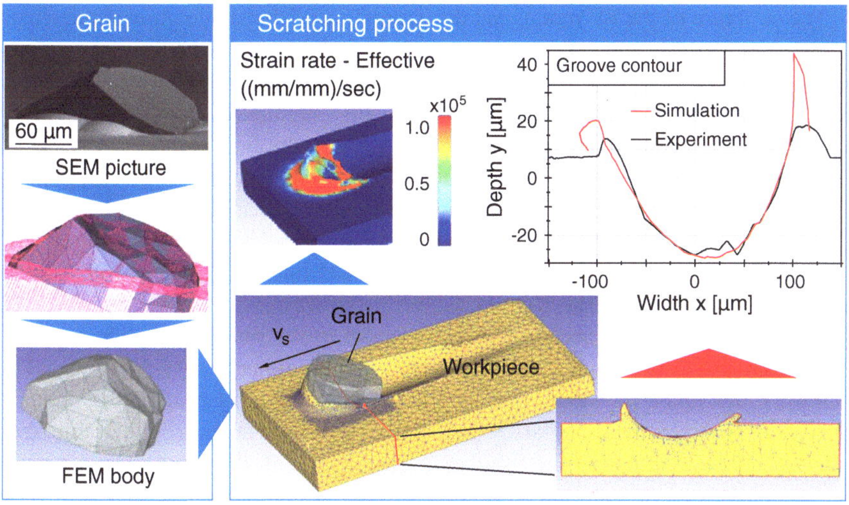

Fig. 4.8 FE simulation of the single grain scratching process

varies both due to the different cutting depths and different grain shapes. As a result, there is no unique correlation between the scratching groove section areas and the contact areas of grain and workpiece. Consequently, the scratching groove section area could be identified as a relevant parameter, though it must be supplemented by other characteristic parameters in ongoing explorations, in order to represent the influence factors that were not considered in the scratching section area.

In experimental studies, potential influencing factors can only be partially adjusted in a controlled and individual manner. Thus, first of all, the grain shape characteristics, such as the rake angle and the section area or the radius at the crest cannot be arbitrarily combined by the use of standard CBN-abrasive grains, but specified by the available grains, and can only be recorded before the test.

For this reason, single grain engagement was also simulated with a thermo-mechanical FE model. This model permits freely variation of the grain shape and exploration of its influences in a controlled manner. The challenge arises from the adaptation of the model to the geometric and kinematic boundary conditions of the grinding process. In particular when grinding in the part material, high temperatures [more than 1,100 °C (Zäh et al. 2009)] and strain rates [10^5–10^7 s^{-1} (Steffens 1983)] occur. On the current test beds, for these conditions, parameters able to represent the material behaviour cannot be obtained adequately. Consequently, material laws based on material data determined in experiments at lower strain rates (up to 10^4 s^{-1}) are used for simulations. These laws are extrapolated into the zone of high strain rates without being verifiable. For this reason, the temperature values measured at defined points in the single grain scratching experiments were applied to calibrate the FE model and thus to enhance the results' accuracy.

The structure of the FE model and the simulation of the single grain engagement was conducted by means of the software Deform 3D in cooperation with partners from the subproject A01 and the Fraunhofer Institute for Machine Tools and Forming Technology (IWU) in Chemnitz. In this simulation, the geometry of an actual abrasive grain was used, which was also applied to the single grain scratching experiment mentioned before. The grain geometry was mapped by means of three scanning electron microscope (SEM) images, captured from different perspectives, found by means of a triangulation technique in the form of a 3D point cloud. Afterwards, this cloud was transferred into a Finite Element solid in several steps (see Fig. 4.8).

In the simulation, the abrasive grain was used in the same orientation as in the experimental studies and at an identical circumferential speed of $v_s = 10$ m/s. The Johnson-Cook material model was applied to the representation of the material behaviour, whereby the material specific parameters were taken from literature (see Fig. 4.9). When comparing the simulation outcomes with the experimental results, the scratched groove contours were seen to strongly coincide (see Fig. 4.8, right top). Minor deviations were found in the lower zone of the contour, resulting from the transformation of the grain geometry and the simplification of geometry performed. Lesser differences at the contour's boundary are due to errors from tactile measurement of the scratching groove generated in experiment, and, in turn, resulting from the geometry of the measuring needle used. However, the resultant process forces are in simulation only about half of the force values measured in experiments. In simulation, high strain rates of $\dot{\varepsilon} > 10^5\ s^{-1}$ can be determined (see Fig. 4.8, middle top). When considering the material constants (see Fig. 4.9), it becomes obvious that, in the case of the parameters used, the influence of the strain rate (constant C) is small. This is one potential reason that the resultant force values are too low. Consequently, the material constants applied to the Johnson-Cook model, and further parameters with effects outside this material law in terms of the scratching forces, such as the coefficient of friction, have to be checked and adapted. To do this, the results gained from investigations still to be carried out on a Split-Hopkinson bar test bed will be referred to.

$$\sigma = [A + B \cdot \varepsilon^n] \cdot \left[1 + C \cdot ln\left(\frac{\dot{\varepsilon}}{\dot{\varepsilon}_0}\right)\right] \cdot \left[1 - \left(\frac{\theta_W - \theta_0}{\theta_m - \theta_0}\right)^m\right]$$

A (MPa)	B (MPa)	n	C	m
688.17	150.82	0.3362	0.04279	2.7786

Fig. 4.9 Johnson-Cook material law equation and constants for 100Cr6—AISI 52100 (Ramesh and Melkote 2008)

4.3.3 *Transferring the Energy Model onto the Multi Grain Engagement*

Having created the energy model for single grain engagement, this model has to be transferred to multi grain engagement and thus also to the grinding process. For this purpose, multi grain scratching experiments are conducted, following a strategy analogous to those of the single grain scratch tests. Wheels equipped with several abrasive grains are brought into engagement with the workpiece. In the process, the forces that appear and the temperatures are recorded, and the energy is concluded from the engagement geometry. The abrasive layer topography is represented by the characterising parameters, identified during the single grain scratching experiments and mapping the influence of the grain shape onto heat generation. These characteristic parameters are an input variable for the energy model in the form of statistical frequency distributions for the abrasive layer topography used. The influence of different distributions of the grains and the grain shape properties is analysed by using different multi-grain scratching tools. Apart from the existing abrasive layer topography, the geometric-kinematic process parameters determine which zones of topography and thereby which grains engage with the workpiece. Accordingly, the distribution of the grain shape's characteristic parameters also depends on the process parameters. To consider this relationship, the energy model is connected with a grain engagement model (see Fig. 4.2). This model uses the limiting angle of cutting edge offset $\varepsilon_{\mathrm{Grenz}}$, representing the shadowing effect of the single grains on one other as a function of the process parameters (Kassen 1969), see (4.3).

$$\tan \varepsilon_{Grenz} = \frac{v_w}{v_s} \cdot 2 \cdot \left(\frac{a_e}{d_s}\right)^{\frac{1}{2}} \tag{4.3}$$

v_w workpiece speed
d_s grinding wheel diameter
v_s grinding wheel circumferential speed
a_e infeed

The grain engagement model determines, based on $\varepsilon_{\mathrm{Grenz}}$, the zones of the existing abrasive layer topography that are engaged (see Fig. 4.10) (Duscha et al. 2009).

From these zones, the characterisation parameters, such as the contact surface in tangential or normal direction, can be abstracted. Based on the outcomes of the single grain scratch tests, the grain engagement model can be continuously extended in order to obtain the relevant characteristic parameters of the abrasive layer topography.

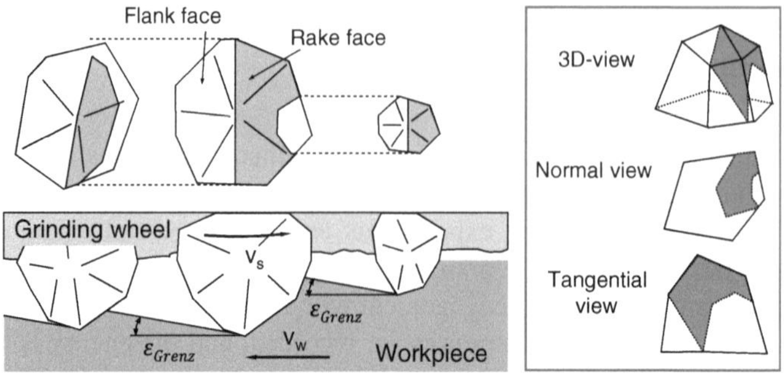

Fig. 4.10 Qualitative representation of the limiting angle of cutting edge offset and the surfaces engaged during machining (Duscha et al. 2009)

4.4 Classification of Outcomes in the CRC/TR 96

The metal removal process is a major source of heat in the machine tool. In grinding, in particular, a vast amount of the mechanical energy introduced is converted into heat due to the many abrasive grains interacting with the workpiece, which are characterised by negative rake angles. According to that circumstance, the energy model for the grinding process is a fundamental prerequisite for the real process representation of the machine tool to be developed in the CRC/TR 96. Only if the thermal energy generated by the grinding process is considered the compensation solutions can be designed according to demand and the correction solutions can be sufficiently validated.

4.5 Outlook

Modelling of the heat source in the grinding process as a function of the process parameters is the basis for another issue in further research projects—modelling of the heat dissipated in the components involved in grinding. For this task, the paradigm of the inverse determination of heat fluxes is to be applied, as it already has been used in subproject B02. Based on the calculations of the heat sources previously performed, temperatures are to be determined in experiment in a discrete local and temporal resolution in order to draw conclusions about thermal conduction and thereby the heat fluxes. The heat fluxes, in turn, are significant input variables for the subprojects A01, A04 and B02 (see Chaps. 2, 4 and 11) in the consideration of the cutting tools, fluid systems, as well as the heat flux across machine components.

References

Brinksmeier E (1982) Randzonenanalyse geschliffener Werkstücke. Dissertation, University of Hannover

Brinksmeier E, Aurich CJ, Govekar E, Heinzel C, Hoffmeister H-W, Peters J, Rentsch R, Stephenson DJ, Uhlmann E, Weinert K, Wittmann M (2006) Advances in modeling and simulation of grinding processes. Ann CIRP 55(2): 667–696

Denkena B, Köhler J, Kästner J (2012) Chip formation in grinding: an experimental study. Prod Eng 6(2):107–115

Duscha M, Klocke F, Wegner H, Gröning H (2009) Erfassung und Charakterisierung der Schleifscheibentopographie für die anwendungsgerechte Prozessauslegung. Teil 2. Diam Bus 28(2):28–33

Kassen G (1969) Beschreibung der elementaren Kinematik des Schleifvorganges. Dissertation, RWTH Aachen University

Klocke F, König W (2005) Fertigungsverfahren Band 2: Schleifen, Honen, Läppen. Springer, Berlin

Malkin S, Guo C (2008) Grinding technology. Theory and applications of machining with abrasives, 2nd edn. Industrial Press, New York

Marinescu ID (2004) Tribology of abrasive machining processes. Andrew, Norwich

Ramesh A, Melkote SN (2008) Modeling of white layer formation under thermally dominant conditions in orthogonal machining of hardened AISI 52100 steel. J Mach Tools Manuf 48:402–414

Rasim M, Duscha M, Klocke F (2013) Innovative Versuchsmethodik zur Identifikation der thermischen & mechanischen Werkstoffbeanspruchung während der Spanbildungsphasen beim Schleifen. In: Hoffmeister H-W, Denkena B (eds) Jahrbuch Schleifen, Honen, Läppen und Polieren, vol 66. Vulkan, Essen, pp 2–19

Rasim M, Klocke F, Weiß M (2014) Identifikation der Spanbildungsphasen beim Schleifen. In: Neugebauer R, Drossel W-G (ed) Innovations of sustainable production for green mobility energy-efficient technologies in production part 1. 3rd international Chemnitz manufacturing colloquium ICMC 2014 3rd international colloquium of the cluster of excellence eniPROD reports from the IWU, vol 80, pp 813–831

Steffens K (1983) Thermomechanik des Schleifens. Dissertation, RWTH Aachen University

Tönshoff HK, Inasaki I, Paul T, Peters J (1992) Modelling and simulation of grinding processes. Ann CIRP 41(2):677–688

Zäh MF, Föckerer T, Brinksmeier E, Heinzel C, Huntemann J-W (2009) Experimentelle und numerische Bestimmung der Einhärtetiefe beim Schleifhärten. Einflüsse und Absicherung der Wärmequellenmodellierung. wt Werkstattstechnik online 99(1–2):49–55

Zeppenfeld C (2005) Schnellhubschleifen von [gamma]-Titanaluminiden. Dissertation, RWTH Aachen University

Chapter 5
Thermo-energetic Modelling of Fluid Power Systems

Juliane Weber and Jürgen Weber

Abstract The subproject A04 is concerned with the analysis of fluidic overall systems and components in machine tools are investigated. The target of the research activities is the development and validation of calculation models that represent the thermo-energetic behaviour of fluid systems. Based on experiments carried out on an overall system, it is possible to identify the key fluid power components and abstract fundamental statements about their working structure and functionality. The outcomes of these investigations form the basis for the engineering of appropriate calculation models. This paper explains the methodological approach in detail. The initial outcomes of the experimental analysis of a milling centre are introduced. Four essential fluidic subsystems are identified and investigated. The work or motor spindle is regarded as a relevant key component in fluidic tempering systems. The calculation models and simulation results, as well as the experimental studies to validate the models, are introduced specifically for the stator cooling unit of a motor spindle.

5.1 Introduction

In machine tools, a huge variety of fluidic components and systems is used—these include, for instance, pumps, heat exchangers or cooling components in motors and work spindles. While the drive tasks for feed motions and tool or part clamping are receding further and further into the background (Kuttkat 2011; Kraus 2012), tempering of the machine tool is becoming more and more important. Fluid power systems can be used to cool or preheat both individual components and entire

J. Weber (✉) · J. Weber
Faculty of Mechanical Science and Engineering, Institute of Fluid Power,
Technical University Dresden, Dresden, Germany
e-mail: juweber@ifd.mw.tu-dresden.de

J. Weber
e-mail: weber@ifd.mw.tu-dresden.de

K. Großmann (ed.), *Thermo-energetic Design of Machine Tools*,
Lecture Notes in Production Engineering, DOI 10.1007/978-3-319-12625-8_5

assemblies (e.g. frame structures) and significantly affect the thermo-elastic characteristics of machine tools.

However, increasing complexity and performance of the fluid power systems also results in an increase in energy consumption (Denkena 2011), which has to be discussed from economic and ecological perspectives. Up to now, the optimisation problem arising from additional energy demand, on the one hand, and thermal characteristics, on the other hand, could not be solved sufficiently due to the complex system structures, absence of fundamental research and design tools.

The target of research in subproject A04 is the engineering of thermo-energetic calculation models of fluidic tempering systems in machine tools. Following a holistic approach, fundamentals of modelling and calculation methods for a scientifically justified interpretation, with a particular focus on optimal thermal behaviour with minimal use of auxiliary energy, are provided. The system boundaries are defined by the surrounding component wall or the exit plane of the fluid into the environment. Since machine tools for metal cutting are extremely precise, high accuracy is also necessary for the thermo-energetic calculation of the fluidic systems in order to be able to represent the three-dimensional energy exchange operations that cross the boundaries of a component.

Consequently, the refinement of network-based simulation models (Gebert 1997; Bossmanns 1999; Chien 2008) is a main issue of research.

5.2 Approach

The approach is subdivided into four subdomains altogether, as shown in Fig. 5.1. In the first step, the fluid power systems are studied directly on the machine tool in experiments. The goal is to provide basic statements on the working structure and modes of operation have to be elaborated. Based on these outcomes, in a second step, the key components to be studied in experiments have to be chosen and analysed on suitable test stands. These experimental investigations—both of the system as a whole and of the key components—are an important precondition for modelling. Relevant characteristic variables and parameters to be used for the calculation models can be extracted from the measured data. The development of thermo-energetic component models starts with the analysis of available simulation tools as well as the identification of the specific modelling requirements. Afterwards, in step three, comprehensive numerical simulations (esp. CFD simulations) are conducted. Using these analyses, it is possible to visualise the thermo-energetic exchange processes in high resolution both temporally and spatially. They are used as a tool to abstract structures and parameters for engineering of component-based node models. Furthermore, the analyses can be employed—apart from the measured data—as an additional reference for comparison. Unlike numerical CFD models, lumped parameter models allow for quicker calculation with the same or possibly even less calculation power. Consequently, the latter are extremely well suited to represent entire system structures by combining individual component

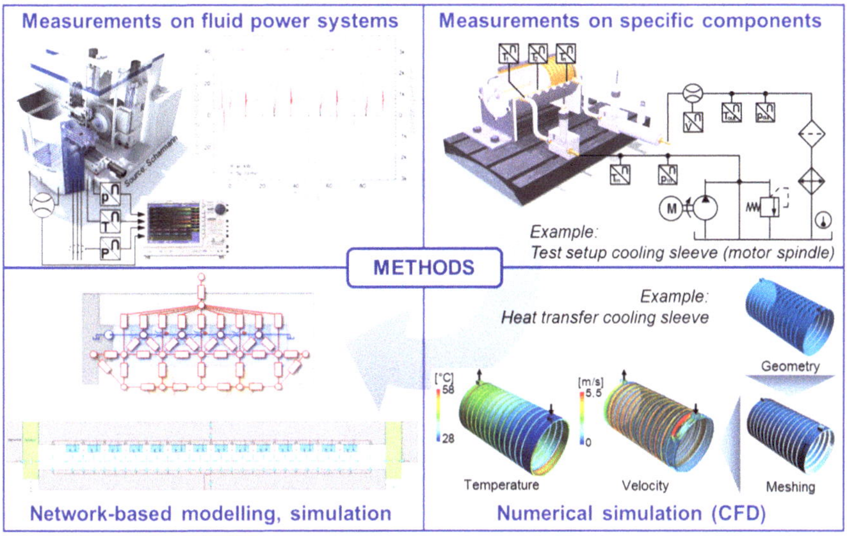

Fig. 5.1 Overview of methodology

models with one another. Thus it is possible to represent the emerging physical relationships and interactions among the components at the system level. Finally, the system behaviour is repeatedly validated by experiments in order to check the extent to which the new component models inside a system model are suitable for precise calculation and ensure that an impermissible number of errors do not occur.

5.3 Results

5.3.1 Complete Machine Analysis of a Milling Centre

Figure 5.2 shows the investigated fluid systems of the demonstrator machine with the related key components. The machine is equipped with five separately driven axes (X-, Y-, Z-, B- and U axes). The headstock (2) can be traversed in Y direction, and the column (1) can be traversed in X direction by means of the driving motors (8) and (9). Rotation (B axis) and horizontal longitudinal motion of the table (4) in Z direction are made possible by another two servo motors. An additional servo motor moves the radial facing slide (U axis, not explicitly illustrated in Fig. 5.2).

Altogether, four fluid systems have to be investigated in detail according to their specific function: The cooling lubricant system, the cooling circuit, the lubrication circuit and the hydraulic system. All of the pumps in the fluid systems are driven by asynchronous motors (three-phase current of 400 V and 50 Hz).

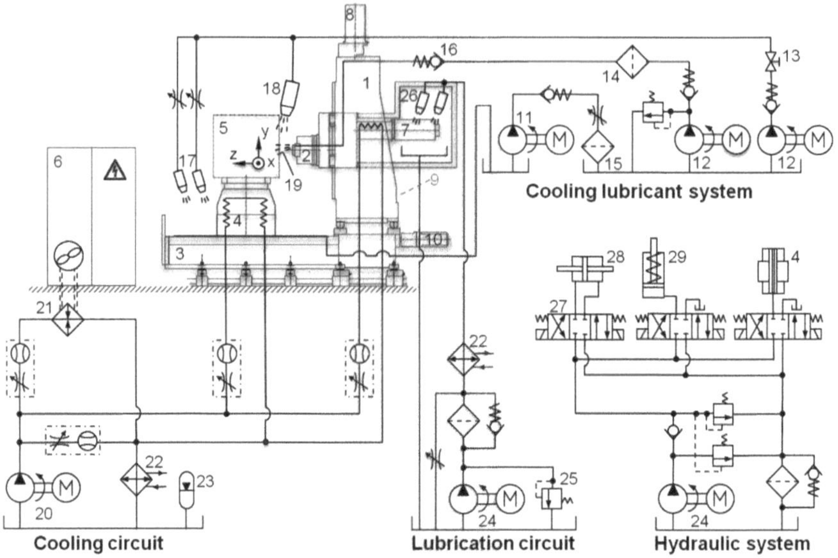

1 Column	9 X axis drive (spindle)	17 Nozzles for flushing	23 Accumulator
2 Headstock	10 Z axis drive (table)	18 Nozzle for external coolant supply	24 Fixed displacement pump
3 Bed	11 Centrifugal pump		25 Pressure relief valves
4 Rotary table	12 Submersible pump	19 Internal coolant supply (through the tool)	26 Nozzles (cooling headstock, lubrication (e.g. main drive)
5 Workpiece	13 Shut off valve		
6 Electrical cabinet	14 In-line filter	20 Cooling unit	27 Directional control valves
7 Main drive	15 Belt filter	21 Air-cooling cabinet	28 Tool (clamping/release)
8 Y axis drive (spindle)	16 Check valve	22 Coolant device	29 Break (at standstill)

Fig. 5.2 Simplified representation of the relevant fluid power systems in the machine tool investigated (Weber 2014)

In the cooling lubricant system, two submerged pumps (12) supply (feed) the metalworking fluid to the machining zone. This is necessary in order to reduce friction between workpiece and tool and thus to diminish heat generation, as well as to guarantee heat and chip removal out of the machining zone. The high pressure pump is used for internal coolant supply through the tool (19). To avoid contamination, a filter element (14) is integrated here. Furthermore, the circuit is prestressed to a pressure of 1 bar by means of the check valve (16). The low pressure pump is designed for flushing (17) of the machining zone and external coolant supply (18). The contaminated fluid is fed into a drip container by means of a chip conveyor. In the next step, the contaminated fluid is recycled by flushing it onto a belt filter (15) by a centrifugal pump (11) and thus separating it from the chips.

The hydraulic system is used for the clamping of tool (28) and rotary table (4), as well as to actuate the retaining brake (29) of the headstock. A fixed displacement pump (24) is integrated in the hydraulic unit. The individual components can be clamped or unclamped by pressurization of each feeding line by means of directional control valves. The current states of each component are supervised by means of additional pressure switches.

In the cooling circuit, a fixed displacement pump (20) is used as drive. To cool the fluid to a predefined initial temperature, a heat exchanger is additionally integrated into the circuit. For use as an antifreeze, as well as to avoid corrosion and furring, a mix of water and max. 20 % Antifrogen® N is used as fluid. The work spindle drive (7), rotary table (4), and the electrical cabinet (6) have to be cooled.

To estimate the power consumed by the machine tool's fluid systems, it is necessary to distinguish different operation states. They can be classified into machining times and auxiliary times according to Götze (2010). Four operation states in terms of machine use were abstracted for the investigations conducted here:

- Warm-up phase (commissioning of the machine and data capturing until a quasi-stationary state of the fluid systems has been established),
- Setup mode (tool/part change and CNC programming),
- Manufacturing process (layer-by-layer plane milling using two different cutting tools),
- Idle run (variation of spindle speed, cooling lubricant (CL) supply and axis position).

Power values were captured for the different operation states of the milling machine in order to estimate energy requirement (power consumption). Figure 5.3 shows the required power as determined from the measured data over a defined test cycle in idle run mode, which was explained in detail in Weber (2014). For this test cycle, lasting as a whole t = 275 s, an overall (electrical) power consumption of approximately 3,300 kJ was measured. The largest percentage of the energy consumption, about 45 %, emerges when driving the axes. The fluidic systems demand

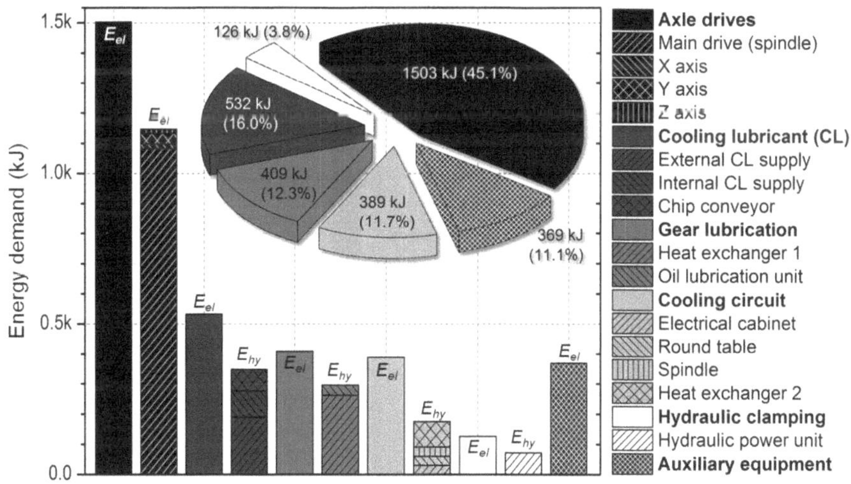

Fig. 5.3 Power consumption of the studied subsystems and components in idle run (test cycle lasting t = 275 s)

an overall ratio of approximately 44 %. The remaining 11 % is required for auxiliary consumers, such as CNC, lightening and pneumatic systems. The cooling lubricant system, demanding approximately 16 % of the consumed power, represents the maximal percentage among the fluid systems. Power consumption of the cooling system is approximately 12 % and corresponds roughly with the percentage consumed by the lubrication system. The least amount of power is consumed by the hydraulic clamping system with approximately 4 %.

Figure 5.4 illustrates the energy flow for the cooling circuit, determined from the measured temperature, pressure and volumetric flow values. Energy flow can be differentiated into static, dynamic and thermal energy in the cooling agent. Thermal energy is calculated using a reference level of T_{ref} = 24 °C (as a rule, it corresponds

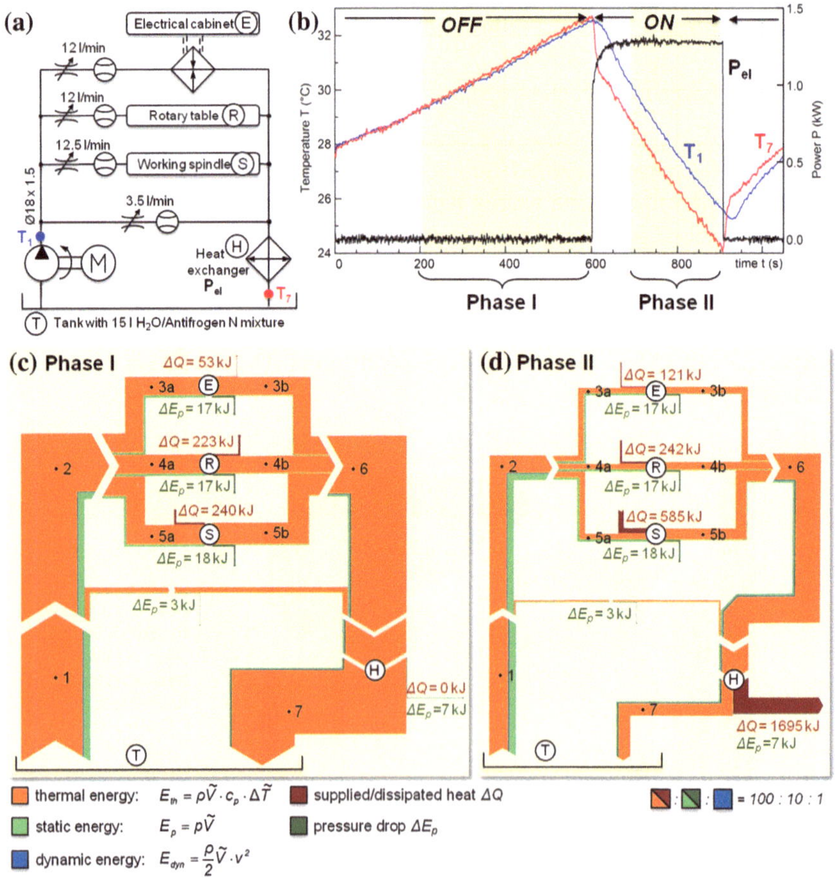

Fig. 5.4 Cooling circuit in idle run mode. **a** Simplified circuit diagram, **b** fluid temperature curve T_1 (immediately behind the cooling pump) and T_7 (immediately behind the heat exchanger), electrical power characteristics P_{el} of the heat exchanger, **c** Sankey diagram for the inactive phase of the heat exchanger, **d** Sankey diagram for the active phase of the heat exchanger (Weber 2014)

to the ambient temperature). Concerning the operation mode of the heat exchanger in the cooling system (two-position controller), in turn, two different phases have to be distinguished (illustrated in Fig. 5.4b). During the first phase, the heat exchanger is switched off, and the cooling agent bypassing it warms up. As soon as a defined maximal temperature has been achieved, the heat exchanger in the second phase is switched on. This switch-on phase lasts until the fluid has reached a defined lower temperature limit. For the analysis of energy flux, it is sufficient to consider the steady state of each phase, which means the temperature changes have to be constant. As the energy flux diagram shows, the fluid cools down in the inactive phase of the heat exchanger (Fig. 5.4c) in the electrical cabinet (E) and the rotary table (R). Fluid is only warmed up in the spindle (S), or, in other words, only the spindle is cooled during the inactive phase of the heat exchanger. The fluid is cold enough to cool even the electrical cabinet and the rotary table only in the active phase (Fig. 5.4d). As can be seen in the two energy flux diagrams, the largest amount of heat is entered into the cooling circuit through the work spindle. For this reason, the work spindle is a relevant main component that has to be explored in detail.

Table 5.1 summarises the share of the energy consumed that relates to each position for the Sankey diagrams shown in Fig. 5.4. The different shares of the energy usage are calculated according to the equations listed in Fig. 5.4 for each operation phase. It can be seen that share of energy consumed as thermal energy is higher by many orders of magnitude than that of the static or dynamic energy. Consequently it makes sense to further analyse the thermal effect in detail separately.

Table 5.1 Share of energy consumed (static energy E_p, dynamic energy E_{dyn}, thermal energy E_{th}) according to the positions enumerated in Fig. 5.4

Position \ Energy	E_p (kJ)	E_{kin} (kJ)	$E_{th.I}$ (kJ) *(cf. Fig. 5.4c)*
1	78.0	1.04	6,701
2	71.0	0.79	6,115
3a	23.4	0.03	2,010
3b	6.4	0.03	1,957
4a	23.4	0.03	2,034
4b	6.4	0.03	1,810
5a	24.4	0.03	2,071
5b	6.7	0.03	2,311
6	19.5	1.04	6,078
7	14.3	1.04	6,828

5.3.2 *Experimental Component Analysis of the Motor Spindle Cooling Sleeve*

In the subproject, a test bed was developed (Fig. 5.5) that made it possible to capture the fluid and solid temperatures in the stator cooling sleeve under defined boundary conditions and to process them using a computer. At the same time, the pressure at the entry and exit, as well as volumetric flow, can be measured. Thanks to the modular test bed structure, it is possible to exchange the cooling sleeve easily and thereby investigate different flow geometries in detail. The stator was replaced with two heating cartridges that were centrically inserted into an aluminium cylinder (the so-called heating element). Energy is input by means of two heating cartridges whose temperature or power can be controlled autonomously. Temperature is captured points-wise not only in the fluid, but also on the internal wall of the cooling sleeve and the housing surface. For this strategy, thermocouples are placed on the corresponding components in axial direction with defined distances. Variable parameters are the heat amount entered, volumetric flow and the composition of the cooling liquid. Additionally, a thermographic camera is used for areal capturing of temperature and characterisation of the temperature distributions on the housing surface.

The first tests were conducted using water as the cooling agent. A constant entry temperature of 20 °C can be adjusted by means of a temperature sensor integrated into the cooling unit. For fluid temperature measurement, the thermocouples were electrically insulated with a thin lacquer layer and each was attached to a PVC screw. If these are screwed into the housing, the test probe protrudes into the fluid flow in the helical channel, and the fluid temperature can be captured. Thermocouples of the T type, whose measuring uncertainty is ±0.5 K due to the class 1 calibration conducted, are employed as temperature sensors. A turbine flow sensor of ±0.5 %FS (full scale) accuracy is used to record the volumetric flow. The pressure in forward and return flows is also measured. Accuracy of the installed pressure-measuring transducer is inside ±0.123 %FS. All signals are captured by means of an M series' NI multifunctional board and processed.

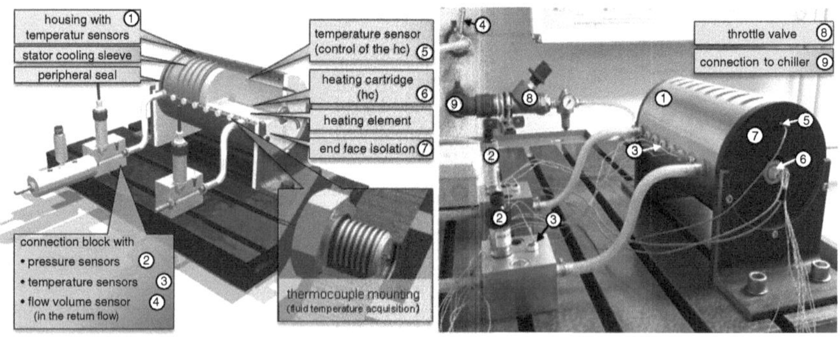

Fig. 5.5 Test bed CAD model (*left*) and practice implementation (*right*)

5.3.3 Simulation Model for the Calculation of the Motor Spindle Cooling Sleeve

Simulation models with lumped parameters are mainly characterized by comprehensive interfaces with external CAx and database programs, short calculation times and, as a result, cost savings. As can be seen in Fig. 5.6, the node model of the motor spindle cooling system consists of individual units describing the physical relationships of the system model. The potentials required for calculation are transmitted via connections between the individual units. The complete model is composed of individual simple partial elements whose characteristics can be described by equations established in thermodynamics and fluid technology (Sigloch 2009; VDI 2006; Schmidt 1967; Hak 1956).

Warming of the fluid along the spindle axis is illustrated for different volumetric flows or different heating element temperatures in Fig. 5.7. Here, the numerical and network-based simulation results, as well as the temperature values in the fluid found in experiments, are compared. As expected, fluid temperature decreases as volumetric flow increases or heating element temperature decreases. The maximal temperature in the exit zone is achieved due to continuous heat absorption by the fluid along the spindle axis. As can be seen, for the onefold helical stator cooling sleeve, the node model developed already corresponds closely to the numerical simulation and the data obtained in experiments. Here, the two upper diagrams indicate the relative deviation between the node model (SimX) and the numerical simulation (CFD) or the experiment (Exp.). In comparison with the values determined in experiments, the deviation is maximal 2.4 % (Fig. 5.7a) or 2.8 % (Fig. 5.7b).

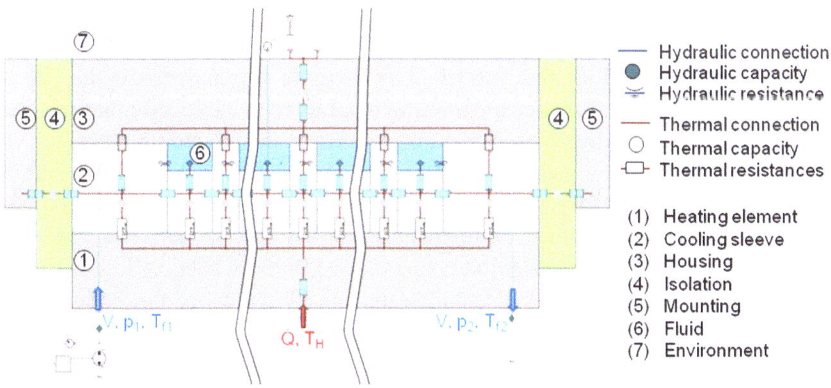

Fig. 5.6 Node model of the onefold helical stator cooling socket in the simulation ambient ITI SimulationX (*abbrev.* SimX), according to Weber (2013)

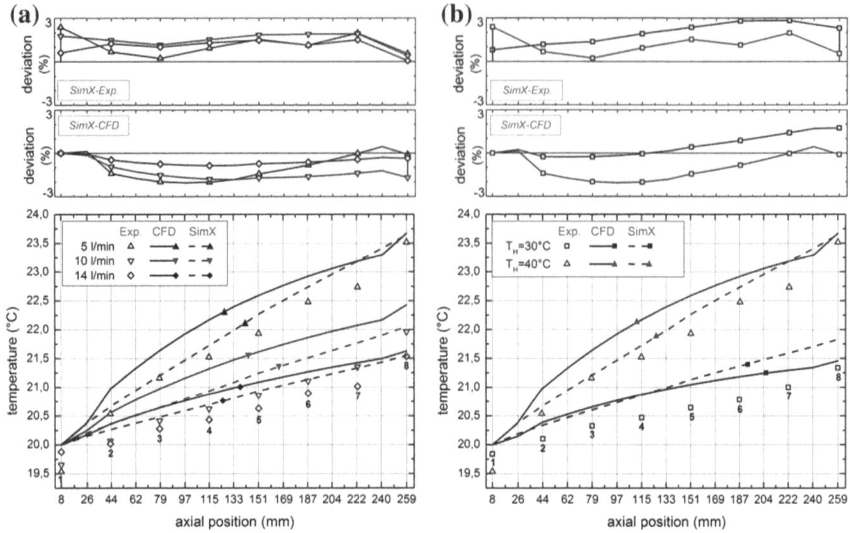

Fig. 5.7 Simulation outcomes for fluid warming along the spindle axis; **a** for a constant heating element temperature T_H = 40 °C at different volumetric flows; **b** for a constant volumetric flow $\dot{V}$ = 5 l/min at different heating element temperatures

5.4 Classification of Outcomes CRC/TR 96

In high speed milling, above all, thermal strains resulting from heat entry have a significant effect on manufacturing accuracy and production quality. Hence targeted control of the temperature distribution—most of all in the spindle—is the main precondition for highly precise machining. In subproject A04, a lumped parameter model describing the heating and flow processes emerging in a high speed spindle has been developed. With this model, it is possible to characterise the forced convection in the cooling water in detail, as well as to achieve an enhanced prediction of temperature distribution and an increase in manufacturing accuracy. In comparison with a model using distributed parameters (CFD), the lumped parameter model also benefits from less effort in modelling and shorter calculation times. The combination of individual component-based network models also makes it possible to represent entire system structures (such as the whole cooling circuit).

The resulting network-based calculation models for typical cooling circuits can be integrated into the complete model of the virtual machine tool, which is under development in subproject A05. The model makes it possible to compute flows, pressure drops and heat transfers in the circuits. Thus, it is possible to make available information about the thermal characteristics and heat entry into the machine structure or the active mechanical components for subproject B01 and subproject B03. Predictions of the fluid systems' cooling performance in motor spindles contribute to subproject C04 during motor modelling and optimisation. The numerical CFD models were also analysed and evaluated in cooperation with the subproject A07.

References

Götze U et al (2010) Energetische Bilanzierung und Bewertung von Werkzeugmaschinen. In: 1. Internationales Kolloquium des Spitzentechnologieclusters eniPROD, Chemnitz, pp 157–184, 24–25 June 2010

Bossmanns B, Tu JF (1999) A thermal model for high speed motorized spindles. Int J Mach Tools Manuf 39(9):1345–1366

Chien CH, Jang JY (2008) 3-D numerical and experimental analysis of a built-in motorized high-speed spindle with helical water cooling. Appl Therm Eng 28(17–18):2327–2336

Denkena B et al (2011) Effiziente Fluidtechnik für Werkzeugmaschinen. Werkstatttechnik online 101(5):347–352

Gebert K (1997) Ein Beitrag zur thermischen Modellbildung von schnelldrehenden Motorspindeln. TU Darmstadt, Shaker Verlag, Aachen

Hak J (1956) Lösung eines Wärmequellen-Netzes mit Berücksichtigung der Kühlströme. Archiv für Elektrotechnik 42(3):137–154

Kraus J (2012) Hydraulik fordert Elektromechanik in der Energieeffizienz heraus. Maschinenmarkt (5):32–35

Kuttkat B (2011) Elektromechanische Spannmittel verdrängen energiefressende Fluidik. Maschinenmarkt 3:24–25

Schmidt EF (1967) Wärmeübergang und Druckverlust in Rohrschlangen. Zeitschrift für Technische Chemie, Verfahrenstechnik und Apparatewesen 39(13):781–832

Sigloch H (2009) Technische Fluidmechanik. Springer, Berlin

VDI (ed) (2006) VDI-Wärmeatlas. Berechnungsunterlagen für Druckverlust, Wärme- und Stoffübertragung. Springer, Berlin

Weber J, Weber J (2013) Thermo-energetic analysis and simulation of the fluidic cooling system of motorized high-speed spindles. In: 13th Scandinavian international conference on fluid power, Linköping, Sweden, 3–5 June 2013

Weber J, Weber J (2014) Thermo-energetic analysis of the fluidic cooling systems in tooling machines. In: 9th international fluid power conference, Aachen, 24–26 Mar 2014

Chapter 6
Simulation of Pose- and Process-Dependent Machine Tool Models

Marian Partzsch and Michael Beitelschmidt

Abstract Within the CRC/TR 96, when calculating the thermal behaviour of machine tools, the consideration of the structural variability of machine tools is an approach to enhance the representational quality of the associated simulation models. A potential solution, which uses embedding moving model elements in sequential thermo-elastic analyses using discrete moving loads, is described in the following. The method is implemented in the proprietary FE software *ANSYS*. Based on selected theoretical fundamentals, embedding of motion is outlined in detail, focussing on an efficient use of the modelling tools made available by *ANSYS*. Various non-trivial special features that may hinder a successful simulation are highlighted. Measures to avoid these problems are described. Finally the influence of the method on the simulation results is described using the example of experiments carried out in an application.

6.1 Introduction

A central objective in research is the in-process overall system simulation of the machine tool and test setups investigated in the CRC/TR 96 by means of proprietary FE software. For simulation, the *ANSYS-Mechanical* software package (currently v14.5, issued 27 Feb 2014), which is widely used and tested, is employed. Choosing this system and limiting it to exclusively program-internal functions are measures aimed at ensuring the simple application of the research results by third parties.

M. Partzsch (✉) · M. Beitelschmidt
Faculty of Mechanical Engineering, Institute for Solid Mechanics (IFKM), Technical University Dresden, Dresden, Germany
e-mail: marian.partzsch@tu-dresden.de

M. Beitelschmidt
e-mail: michael.beitelschmidt@tu-dresden.de

K. Großmann (ed.), *Thermo-energetic Design of Machine Tools*,
Lecture Notes in Production Engineering, DOI 10.1007/978-3-319-12625-8_6

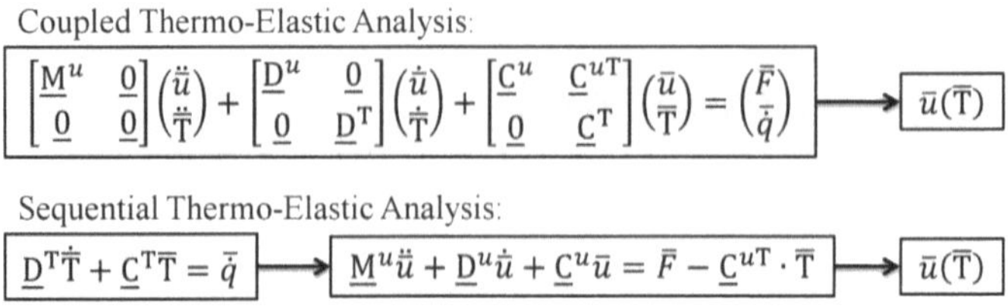

Fig. 6.1 Theory of coupled and sequential analysis for discrete thermo-elastic field problems (common nomenclature)

Based on the requirement for in-process-suitability, another goal is to consider in the model the structural variabilities relevant for machine tools. For this reason, initial investigations deal with the methodology of embedding moving model elements in *ANSYS* in general. The following explanations describe the chosen methodology and the outcomes obtained. First, how to choose the fundamental analysis type is shown.

For the calculation of the thermo-elastic system behaviour, ANSYS provides several coupled-field elements (such as SOLID226) that enable strong coupling of the thermal and structural engineering field problems. To bypass the substantial effort required to resolve the systems by means of asymmetric matrices, a sequential analysis of the thermo-elastic problem is instead conducted (see Fig. 6.1). First, the exclusively thermal field is calculated in-process including the kinematics. With the thermal loads obtained this way, the resultant displacement of the structure due to temperature is subsequently determined at discrete points in time. In this process, the influence of the structural changes on the thermal field via the effects of piezocaloric damping or very large deformations is ignored. This simplification is a reliable assumption for the investigated systems due to the different time constants of the two fields. In the analysis type selected, the exclusive integration of the structural variabilities into the thermal part of the analysis also avoids contact-induced convergence problems, since no unknown structural deformations are generated.

The intrinsic coupling of the fields in *ANSYS* is a standard function in terms of FE technology and thus is not described here. Moreover, structure variability does not affect it, so the problem can be limited to embedding of motions in purely thermal simulations.

6.2 Approach to Mapping of Structural Variability

To implement structural variability in thermal analyses, the functions available in *ANSYS* are combined. Table 6.1 shows which main platforms of the software package are employed in the individual analysis sections, and which work steps are thus particularly supported.

Table 6.1 Supported functions of *ANSYS Mechanical* Capabilities

	ANSYS workbench	ANSYS classic
Analysis section	Pre-processing	Solution
		Post-processing
Facilitated work steps	Simplification of geometry	Definition of constraints
	Mesh control/meshing	Control of analysis
		Memory and administration of results

This methodology is also useful because in the *ANSYS-Workbench*, which is similar to a CAD program, it is possible to approach the structure of the existing problem more heuristically. Subsequent processing in *ANSYS-Classic*—as explained in detail below—is necessary for analyses of variable structures, on the one hand. On the other hand, it simultaneously enables the execution of exact and reproducible simulations due to the implementation and application of APDL scripts.

6.2.1 General Strategy to Represent Discrete Motions

As illustrated in Fig. 6.2, a moving body is embedded during a transient analysis by discrete rigid body displacement via adaptation of the corresponding Boundary condition (BC) at the structural degrees of freedom (DoF) **u**. Their necessary existence during simulation compels meshing of the moving assemblies with coupled-field elements even in exclusively thermal analyses, whereby the field coupling is suppressed by a thermal coefficient of expansion equal to zero.

The aim of the investigations was to integrate the physical interaction of stationary and moving assemblies into simulation. Since defined translational kinematic profiles are mapped, the instantaneous model structure is known at all points during the analysis, so that the heat flux caused by friction $\dot{q}_{fric}$ can be manually applied as the NEUMANN-BC to the contact zone. Apart from this, the model structure is identified automatically [*Pinball* algorithm, see Rust (2011)] by means of the element technology oriented to contact mechanics (such as CONTA174), which later defines the heat flux[1] $\dot{q}_{cond}$ exchanged by conduction by means of the temperature conditions appearing in the contact zone.

After the solids involved in the contact are loaded with the additionally overlapping contact boundary, they are subsequently evaluated as separate thermal marginal value problems. The requirement that the solution at $\Delta t \rightarrow 0$ must converge to the case of a continuous motion is fulfilled by this method.

[1] Corresponds to an individual ROBIN-BC resulting from the surface temperatures involved and the heat transfer coefficient in contact of α_{TCC} where $\dot{q}_{cond} = \alpha_{TCC}(T_C - T_T)$.

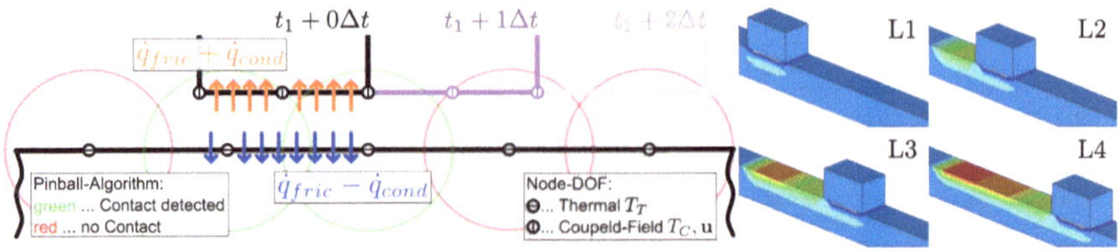

Fig. 6.2 Motion in numerical analyses—*left* schematic illustration of applied theory [from Haber and Partzsch (2013)]—*right* discrete adjustment of displacement for load steps L1 to L4 (Partzsch and Beitelschmidt 2014)

6.2.2 Motion in FE Models—Selected Aspects

The consideration of structural variability must already been taken into account for the modelling process. Thereby the main focus in adapting the pre-processing is on regular meshing of the rail regions that are in contact in the direction of movement. Tetrahedron elements have to be avoided since they result in considerable numerically caused discontinuities in the temperature field that appears, particularly when large time intervals are chosen. Instead, it is recommended to mesh the rails by means of the *Sweep*-method using hexahedrons to allow for a homogeneous rail load orthogonal to the direction of movement.

Figure 6.3 left depicts the execution of this methodology by means of the profile rail guidance investigated in (Haber and Partzsch 2013). The mesh control is mainly conducted through meshing of the basic area and should be performed according to the physical requirements of the system to be simulated. For the case under consideration, this is expressed by a very fine meshing of the thermally active contact regions, meaning the positions where the rolling elements roll along the rail.

As a rule, it is not possible to design the overall FE model in a way that a *Sweep* meshing can be carried out throughout the geometry. This case, as well as the measures recommended as remedies, can be seen in Fig. 6.3, right. Figure 6.3

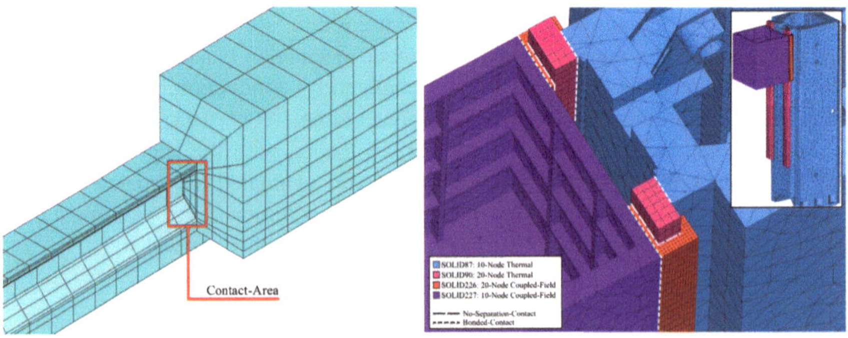

Fig. 6.3 Meshing in FE modeling with structural variability—*left* profile rail guidance (Haber and Partzsch 2013)—*right* column assembly

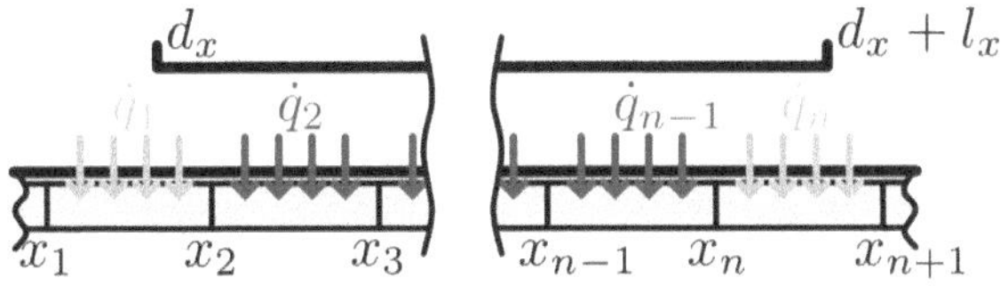

Fig. 6.4 Definition of frictional heat flux on spatially discretised stationary body

outlines an example of an FE model of variable structure—the column assembly (see Chap. 1), which has already been explored repeatedly. For this example, in the *ANSYS DesignModeler,* each of the two assemblies was separated into a *sweep*-able body and a body that could be freely meshed. Afterwards, their associated FE models can be unproblematically recombined into a thermal unit via a bonded contact.

At present, the Coupled-Field elements required for the motion are only available in *ANSYS Classic*, so that it is still impossible to completely process items with structural variability in the *ANSYS Workbench.* That means that the change of the platform that is performed after meshing, has even until now been constrained by the limitations of the software. As previously mentioned in Table 6.1, in this way, among others, the definition of the motion-dependent boundary conditions can be handled more beneficially.

The contact friction, which, in the case of a purely thermal analysis, is interpreted as the NEUMANN-BC $\dot{q}_{fric}$ on the covered rail surfaces (compare Fig. 6.2), is in many simulations an extremely dominant heat source in the systems with structural variability considered here. Their correct mapping is of crucial importance for the quality of the overall analysis. When positioning the moving assembly arbitrarily in the current load step, in general, there are element layers of the stationary assembly under their ends in the direction of movement, which are only partially covered (see Fig. 6.4). For these layers, heat flux has to be modified according to the overlap per linear scaling. The following heat fluxes $\dot{q}_i$ are yielded for a body of length l_x moved in parallel to the x axis with the current rigid body displacement d_x for the geometric position data from Fig. 6.4 for the covered element layers $i = 1 \ldots n$:

$$\dot{q}_1 = \frac{x_2 - d_x}{x_2 - x_1}\dot{q}_{fric} \quad \dot{q}_{2 \ldots n-1} = \dot{q}_{fric} \quad \dot{q}_n = \frac{d_x + l_x - x_n}{x_{n+1} - x_n}\dot{q}_{fric} \tag{6.1}$$

The necessity of this correction does not depend on the time increment; it is also applicable for analyses close to time convergence. Consequently, it is necessary to take it into account for the analysis method in terms of structural variability.[2]

[2] Disregarding the incorrect heat flux densities in the event that they are not corrected is only conceivable for rail models of high mesh density (h-convergency).

6.3 Results

6.3.1 Example of Profile Rail Guidance

Since the intrinsic results of the methodological explorations have already been explained in the preceding section, here their application should be elucidated using a concrete example. Thus it is to be shown how the motion really impacts the resultant temperature field.

Heat build-up due to friction in a profile rail guidance is investigated. The data for model adjustment and validation were captured in an experiment in the subproject B03 (see Haber and Partzsch 2013). The corresponding FE model was created by means of standardised strategies in thermal analyses and has already been demonstrated (in Fig. 6.3, left).

Figure 6.5 left indicates the measurement and simulation results at two measurement points on the rail for the first 600 s of rail heating due to operation. The carriage moves back and forth at a constant velocity of 20 m/min between the end points of a rail section of 540 mm length. Model adjustment, particularly quantification of BC's, such as moving friction and thermal conduction to the coupling structures, was conducted by means of measurement values for heat build-up over the longer term. The temperature curves as a function of time (see Fig. 6.5) confirm that the model maps the heat build-up on the rail with sufficient accuracy.

In the measurement, the two heating curves shown start with a visible difference, which remains nearly constant during the operation. It is impossible to meaningfully integrate this offset into the simulation.

The real effects on the heat build-up of the structural variability considered here, that is, the oscillation of the rail temperature caused by motion as can be seen in detail in Fig. 6.5 right, is of primary interest indeed. It can be clearly observed that the temperature rises as the carriage passes and then drops down again due to subsequent thermal conduction into the body's interior. Since the heating at first is

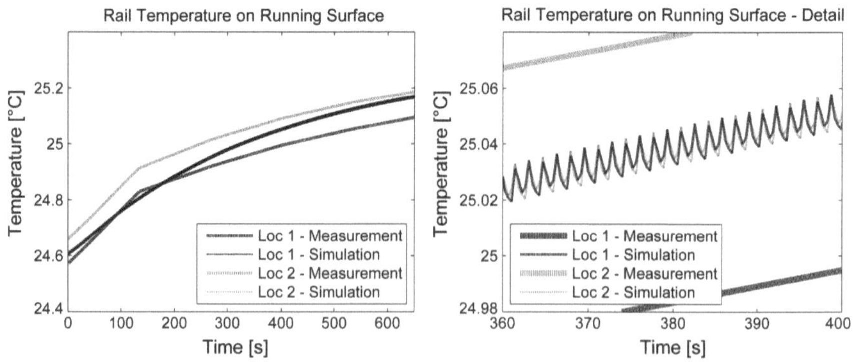

Fig. 6.5 Progress of rail temperature in measurement and simulation—overall period (*left*) and detail (*right*)

greater than the subsequent drop, the system sustains lasting heat, as can be observed in Fig. 6.5, left. The maximal deviations from the assumed average temperature, which result from the motion, are ≈0.01 K and thus very low.

When considering the results, it can be stated that it is not reasonable to consider the motion by the introduced method for the example given. For a model with 20,000 nodes, the comprehensive calculations (1,600 load steps) are not justified given the resulting refinement in the accuracy that can be achieved. Consequently, for the simulation of actual existing systems with friction loads on a similar order of magnitude, in which mainly the long-term heat build-up is of interest, it is preferable to apply load definition methods that permit a more efficient utilisation of large-scale thermal time constants. In an initial approach, it is conceivable, for instance, to use constant friction loads along the overall rail whose height values are provided in proportion to each coverage time per moving period.

A useful application is also conceivable for problems with higher loads and materials with a higher thermal diffusivity. However, in future, the primary use is expected for the reference calculation within the scope of numerical experiments, using models of variable structure, since the method can provide very accurate and reliable solutions for unambiguously defined problems.

6.4 Classification in the CRC/TR 96

As mentioned in Sect. 6.3, at present, the method developed is applied to numerically represent the thermal behaviour of the profile rail guidance [see Haber and Partzsch (2013)].

In addition, in the subprojects A07 [see Chap. 9, (Naumann and Wensch 2013)—*Defect Corrected Averaging*] and A06 [see Chap. 8, (Benner et al. 2013)—*Parametric MOR*], as well as in the subproject A05 [see Chap. 7, (Galant 2012)—*Ordinary MOR on Systems with Structural Variabilities*], some methods at a higher level of mathematical abstraction, which ought to take into account the structural variabilities in numerical analyses with significantly less effort, have been engineered. For their validation, the rather intuitive approach introduced here and its implementation using the trustworthy software ANSYS should provide reliable reference results for different case studies.

6.5 Outlook

Current studies work to refine the technique described here to enhance its efficiency despite the limited range of its application. For this purpose, methods that correct the negative influence of large time intervals on the accuracy of the results are being engineered. The first of these correction methods is described in detail in Partzsch and Beitelschmidt (2014) and counteracts the unavoidable non-uniform distribution of introduced coupling heat, which distinctly affects the result quality in case of large time intervals. Further studies will clarify how a more infrequent updating of the conductive exchange heat affects the results.

In parallel, in continuation of the approaches in Haber and Partzsch (2013) (SOLID structures in relative mutual motion) and Beisitzer and Beitelschmidt (2012) (detailed mapping of the rolling elements by means of LINK elements), future works are dedicated to making possible the intrinsic mapping of the coupling assemblies in FE models with different levels of detail.

References

Beisitzer S, Beitelschmidt M (2012) Fallstudien zur thermischen Simulation von bewegten, profilschienengeführten Baugruppen. Paper presented at the ANSYS conference and 30th CADFEM user's meeting 2012, Kassel, 24–26 Oct 2012

Benner P, Lang N, Saak J (2013) Modeling structural variability in reduced order models of machine tool assembly groups via parametric MOR. PAMM 13:481–482

Galant A (2012) Effiziente Simulation thermo-elastischer Verformungen an Werkzeugmaschinen mit ordnungsreduzierten Modellen bei Berücksichtigung großer Relativbewegungen. Paper presented at the ANSYS conference and 30th CADFEM user's meeting 2012, Kassel, 24–26 Oct 2012

Haber D, Partzsch M (2013) Reib- und Temperaturverhalten von Profilschienenführungen in Experiment und Simulation. Paper presented at the 3rd SFB/TR96-colloquium "Experimentelle Methodik", Aachen, 29–30 Oct 2013

Naumann A, Wensch J (2013) Defect corrected averaging for parabolic PDEs. PAMM 13:509–510

Partzsch M, Beitelschmidt M (2014) An arithmetic correction for avoiding non-uniform heat input distribution caused by translational motions within time-discrete thermal analyses. Proceedings of the ASME 2014 12th biennial conference on engineering systems design and analysis

Rust W (2011) Nichtlineare Finite-Elemente-Berechnungen: Kontakt, Geometrie, Material, 2edn. Vieweg+Teubner, Wiesbaden

Chapter 7
Thermo-Elastic Simulation of Entire Machine Tool

Alexander Galant, Knut Großmann and Andreas Mühl

Abstract This paper presents an integrated approach to simulate the thermo-elastic behaviour of machine tools—from the CAD geometry data to FE modelling up to the thermo-elastic network model of variable structure that can be run in real time mode. The first part outlines the theoretical aspects of the methodology developed, whereas the second part demonstrates implementation in practice by means of concrete examples.

7.1 Introduction

Thermo-elastic models that can be simulated for the overall system consisting of machine tool, process and environment are required for the design of compensation solutions, the evaluation of thermo-elastic behaviour and the correction of thermo-elastic errors during machine tool operation.

If it is important for the *evaluation of the targeted design of the thermo-elastic behaviour* in the design-oriented development phases to be carried out without real-time requirements to be fulfilled by the calculation algorithm, then the modelling, simulation and evaluation process can completely be executed at the FE system level. This approach is shown in the Chap. 6 in the same Lecture Notes.

Minimisation of calculation time is a key requirement for a quick *analysis of a wide spectrum of thermal loads and boundary conditions* and the *correction of thermo-elastic errors* during machine tool operation. Network-based modelling

A. Galant (✉) · K. Großmann · A. Mühl
Faculty of Mechanical Engineering, Institute for Machine Tools and Control Engineering, Technical University Dresden, Dresden, Germany
e-mail: Alexander.galant@tu-dresden.de

K. Großmann
e-mail: knut.grossmann@tu-dresden.de

A. Mühl
e-mail: andreas.muehl@tu-dresden.de

K. Großmann (ed.), *Thermo-energetic Design of Machine Tools*,
Lecture Notes in Production Engineering, DOI 10.1007/978-3-319-12625-8_7

provides a solution: FE modelled machine tool (MT) assemblies without inner relative motions are transformed into compact objects by using model order reduction (MOR) method and are linked into a network model in a form of SIMULINK block diagram.

7.2 Approach

In purely thermo-elastic calculations, it is only necessary to take into account the static deformation effects, since the natural frequencies of the structural dynamics, which are significantly higher, are—as a rule—not excited by relatively slow thermal processes. For this reason, the so-called integration strategy of weak coupling is employed (Groth and Müller 2009; ANSYS 2014): in a first step, the temperature vector $\{T\}$ is determined by means of numerical integration. In a second step, the problem of calculating the deformation vector $\{x\} = f(\{T\})$ is solved as a quasi-static one on a possibly much rougher time scale.

This methodology makes it possible to divide the overall thermo-elastic problem into thermal transient and mechanical quasi-static subtasks. After discretization by means of FE method, both a system of differential Eq. (7.1a) and a system of linear Eq. (7.1b) of the following type are obtained

$$[C_T]\{\dot{T}\} + [K_T]\{T\} = \{\dot{Q}\} \tag{7.1a}$$

$$[K]\{x\} = \{F\} - [K_{xT}]\{T\} \tag{7.1b}$$

with $[C_T]$, $[K_T]$, $[K]$ and $[K_{xT}]$—capacity, conductivity, mechanical stiffness and thermal stiffness matrices, $\{x\}$ and $\{T\}$—the deformation and temperature vectors, $\{F\}$ and $\{\dot{Q}\}$—mechanical and thermal load vectors.

The approach traced here is assembly-oriented. In this context, an assembly is a structural area with no internal relative motion. The assemblies, in turn, are connected to a network model by contact representations mapping the effect of large-sized relative motions like those typical for machine tools.

The associated *modelling and calculation concept* is illustrated in Fig. 7.1. The CAD models of machine structure are detailed in assemblies, from which thermal FE models are generated. The FE models are transformed into objects with substantially fewer degrees of freedom by means of model order reduction methods explained later. The contacts are led back to "external heat sources" mapping the heat exchange depending on the relative motion. Afterwards, the calculated vectors of the reduced degrees of freedom are retransformed into complete real assembly temperature vectors. The pose-specific deformation of the overall structure is finally determined by means of the pose-specific quasi-static Eq. (7.1b).

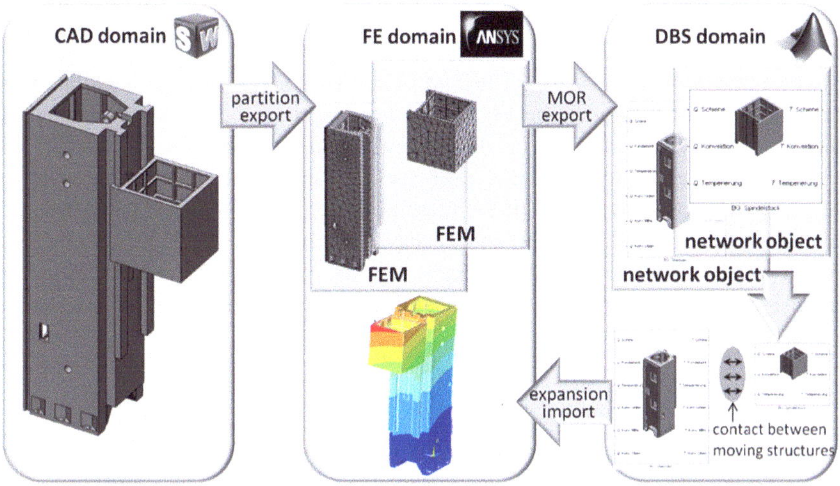

Fig. 7.1 Modeling and calculation concept

7.3 Results

7.3.1 Model Order Reduction

The Model Order Reduction (MOR) methods make it possible to efficiently generate compact model with good approximation properties directly from an original FE model (Schilders et al. 2000; Benner et al. 2005).

The approach starts with a formulation of the thermal subproblem as a state space representation that is entirely equivalent to the linear dynamic system (7.1a):

$$\begin{Bmatrix} u_1 \\ \vdots \\ u_k \end{Bmatrix} \rightarrow \Sigma : \begin{matrix} \{\dot{T}\} + [A]\{T\} & = [B]\{u\} \\ \{y\} & = [C]^T\{T\} \end{matrix} \rightarrow \begin{Bmatrix} y_1 \\ \vdots \\ y_m \end{Bmatrix} \tag{7.2}$$

The MOR method is applied to the thermal subproblem (7.1a) according to the conceptual design of the calculation, Fig. 7.1. For the transformation of the FE model (7.1a) in the form (7.2) see (Großmann et al. 2012b).

The input signals u_i are time-dependent parameters representing the loads applied to the FE model. The temperature vector $\{T\}$ acts as the state vector in system (7.2). The system matrix $[A]$ is determined by the geometry of the assembly to be modelled and its material parameters. Assuming that the thermal system is loaded by convection, the system matrix can be understood as the linear combination

$$[A] = \alpha_1[A_1] + \cdots + \alpha_s[A_s] \tag{7.3}$$

with heat transfer coefficients α_i. The control matrix $[B]$ describes the location of the node positions of the loads the FE system is charged with. The output signals y_i map the state values that can be observed either as temperatures in selected FE nodes or as average temperatures over special ranges. The temperatures to be observed are chosen by means of the measurement matrix $[C]$. If $[C]$ is defined as identity matrix, then this is the special case in which the original system (7.2) provides the overall temperature vector $\{T\}$ as output signal vector $\{y\}$. The dimension N of the overall system Σ is defined by the dimension of the state vector $\{T\}$.

7.3.2 Structure and Parameter Preserving Krylov Model Order Reduction

A special type of MOR method was developed and employed. The basic idea of the method is that the original state vector is approximated in a Krylov subspace by means of its projection (Krylov 1931).

Let $K_n = span\{\{v_1\}, \ldots, \{v_n\}\} \subset \mathbb{R}^N$ be the subspace, and assume that $n \ll N$ and $\{v_1\}, \ldots, \{v_n\}$ its orthonormal base. The matrix $[V_n] = [v_1, \ldots, v_n]$is called the *transformation matrix.*

After insertion of the approximation $[V_n]\{\hat{T}\} = \hat{T}_1\{v_1\} + \cdots + \hat{T}_n\{v_n\}$ into system (7.2) thereby replacing the state vector $\{T\}$ and later multiplication by the orthogonal matrix $[V_n]^T$, the reduced system $\hat{\Sigma}$

$$\begin{Bmatrix} u_1 \\ \vdots \\ u_k \end{Bmatrix} \rightarrow \hat{\Sigma}: \begin{matrix} \{\dot{\hat{T}}\} + [\hat{A}]\{\hat{T}\} & = [\hat{B}]\{u\} \\ \{\hat{y}\} & = [\hat{C}]^T\{\hat{T}\} \end{matrix} \rightarrow \begin{Bmatrix} \hat{y}_1 \\ \vdots \\ \hat{y}_m \end{Bmatrix} \tag{7.4}$$

is generated from the original system Σ. In this equation, $\{\hat{T}\}$ is the vector of the generalized coordinates related to the decomposition of $\{T\}$ according to the base $\{v_1\}, \ldots, \{v_n\}$ of the projection subspace K_n, and $[\hat{A}], [\hat{B}], [\hat{C}]$ defined as follows:

$$\begin{gathered} [\hat{B}] = [V_n]^T \cdot [B]; \quad [\hat{C}] = [V_n]^T \cdot [C]; \; [\hat{A}] = \sum \alpha_l [\hat{A}_l] \\ \text{with } [\hat{A}_l] = [V_n]^T \cdot [A_l] \cdot [V_n] \end{gathered} \tag{7.5}$$

The key task when designing the reduced model $\hat{\Sigma}$ from the original model Σ is to obtain the transformation matrix $[V_n]$ or the Krylov subspace K_n.

The system Σ is linear, so problem (7.2) can be expressed equivalently as given below:

$$\Sigma : \begin{cases} \{\dot{T}_1\} + (\sum \alpha_l [A_l])\{T_1\} = u_1\{b_1\} \quad ; \quad \{y_1\} = [C]^T\{T_1\} \\ \cdots \qquad\qquad\qquad\qquad\qquad\qquad \cdots \\ \{\dot{T}_k\} + (\sum \alpha_l [A_l])\{T_k\} = u_k\{b_k\} \quad ; \quad \{y_k\} = [C]^T\{T_k\} \\ \qquad\qquad \{y\} = \{y_1\} + \cdots \{y_k\} \end{cases} \tag{7.6}$$

where $\{b_1\},\ldots, \{b_k\}$ are the column vectors of the matrix $[B]$.

The k first equations of (7.6) are subsystems of "single input signal" type and can each be handled by means of a classic Arnoldi process (Arnoldi 1951; Lohmann and Salimbahrami 2004; Großmann and Mühl 2010). In this case, the Arnoldi procedure explicitly provides the orthonormal base of the Krylov subspace. It should be mentioned that the parameter vector $\{\alpha\} = \{\alpha_1,\ldots, \alpha_s\}$ is replaced by a 1-vector $\{1,\ldots, 1\}$ to build up the transformation matrices in the Arnoldi process.

$$\begin{array}{lllll} (\sum [A_l]), \{b_1\} : & \overset{Arnoldi}{\longrightarrow} & [V_n^1] & \overset{(5)}{\rightarrow} & [\hat{A}_l^1], \{\hat{b}_1\}, [\hat{C}^1] \\ \cdots \quad \cdots & \cdots & & & \\ (\sum [A_l]), \{b_k\} : & \overset{Arnoldi}{\longrightarrow} & [V_n^k] & \overset{(5)}{\rightarrow} & [\hat{A}_l^k], \{\hat{b}_k\}, [\hat{C}^k] \end{array} \tag{7.7}$$

Consequently, k reduced partial models $(\hat{\Sigma}_1, \ldots, \hat{\Sigma}_k)$ of the dimension $n \ll N$ are generated for k input signals of the original model Σ.

$$\hat{\Sigma} : \begin{cases} \hat{\Sigma}_1 : \left\{\dot{\hat{T}}_1\right\} + (\sum \alpha_l [A_l^1])\{\hat{T}_1\} = u_1\{\hat{b}_1\} \quad ; \quad \{\hat{y}_1\} = \left[\hat{C}^1\right]^T\{\hat{T}_1\} \\ \cdots \qquad\qquad\qquad\qquad\qquad\qquad\qquad \cdots \\ \hat{\Sigma}_k : \left\{\dot{\hat{T}}_k\right\} + (\sum \alpha_l [A_l^1])\{\hat{T}_k\} = u_k\{\hat{b}_k\} \quad ; \quad \{\hat{y}_k\} = \left[\hat{C}^k\right]^T\{\hat{T}_k\} \\ \qquad\qquad \{\hat{y}\} = \{\hat{y}_1\} + \cdots \{\hat{y}_k\} \end{cases} \tag{7.8}$$

The superposition of the output signal vectors of the partial models provides the approximation of the original model's output signal vector (7.2).

Although it is impossible to a priori determine the accuracy of the MOR method, it was possible to successfully verify it in many practical examples. Selected validation experiments are described at the end of this paper.

At the end the following conclusions are derived:

1. The reduced model is of significantly smaller dimension in comparison with the unreduced FE original model if $n \ll N$ and can thus be calculated much more quickly.
2. The generation of the reduced model is independent of the specified input signals u_i and parameter values α_i.
3. The structure of the original system Σ is explicitly preserved with the reduced system $\hat{\Sigma}$.

4. In the special case of the [C] identity matrix, the last equation of (7.8) provides the instructions to retransform the reduced temperature vector to the full temperature vector of the original FE model:

$$\{T\} = \left[V_n^1\right]\left\{\hat{T}_1\right\} + \cdots + \left[V_n^k\right]\left\{\hat{T}_k\right\} \tag{7.9}$$

In the thermal system, the time-dependent load parameters (such as heat flows and ambient temperatures) become a part of the input signal vector $\{u\}$, whereas the convective heat transfer coefficients become a part of the parameter vector $\{\alpha\}$. For this reason, the points 2 and 3 given above mean that the generation of the compact reduced model from the original FE model is independent of the setting of the load parameters and can thus be carried out before parameter setting. Consequently, the one-time generated reduced model can be applied to different load distributions.

The structure preserving Krylov model order reduction method described above is called *MOR-FE method* in the following, since it is carried out immediately after the FE procedure thereby using the FE model as the original model for the MOR process, on the one hand. On the other hand, the outcomes obtained with the reduced calculation can be projected by retransformation to the original FE mesh to be used in, for instance, additional structural analyses in the FE domain or for visualisation purposes (compare Fig. 7.1).

7.3.3 Handling of Structural Variability

7.3.3.1 Thermal Model

Using the MOR-FE analysis described, a small-dimensional thermal model with load parameters standing for the inputs, as well as temperature values for the outputs, can be generated for each assembly (AS). The MOR-FEM-AS objects are the key elements of the network model created further on (compare Fig. 7.5).

The AS objects are coupled with other AS and environmental objects via contact-load conditions. These conditions in particular depend on the relative location of the contacting partners, and can be interpreted as movable heat sources from the view of a single AS object.

Structural variability in the thermal overall model is mapped as follows: at each point in time, the current location of the assemblies in relation to each other and to their ambient space is monitored. Each individual AS model's load parameters depending on this location are calculated and transferred to the AS objects as inputs. Afterwards, state $\left\{\hat{T}_{BG}\right\}$ is calculated on the individual AS models.

7.3.3.2 Mechanical Model

Having obtained the results of the numerical integration of the reduced thermal problem, the thermal state $\{\hat{T}^j\}$ is known for each AS object j at the current point in time. Now we seek the deformation of the machine tool structure consisting of L assemblies in the time interval, in which the AS temperature does not change significantly.

First it is possible to reproduce the temperature vector $\{T^j\}$ for each AS and thus the temperature state $\{T\} = \{\{T^1\}, \ldots, \{T^L\}\}$ of the overall machine tool (MT) structure by means of retransformation (7.9). The deformation $\{x\}$ is found according to Eq. (7.1b) $\{x\} = [K]^{-1}\{F\} - [K]^{-1}[K_{xT}]\{T\}$, whereby the stiffness matrix and the thermal stiffness matrix depend on the assembly location relative to one to another (compare Fig. 7.2). The mutual arrangement of the AS (referred to as *pose* in the following), in turn, depends on the current TCP position in the workspace. As a result, deformation $\{x\}$ is pose-specific.

In the mechanical overall model, structural variability is considered as follows: a regular mesh of TCP supporting positions is placed over the workspace. A defined position of the MT structure exists for each of these discrete supporting positions (from 1 to M). The stiffness matrices $[K^1], \ldots, [K^M]$ and the thermal stiffness matrices $[K^1_{xT}], \ldots, [K^M_{xT}]$ are calculated previously and the deformation vectors

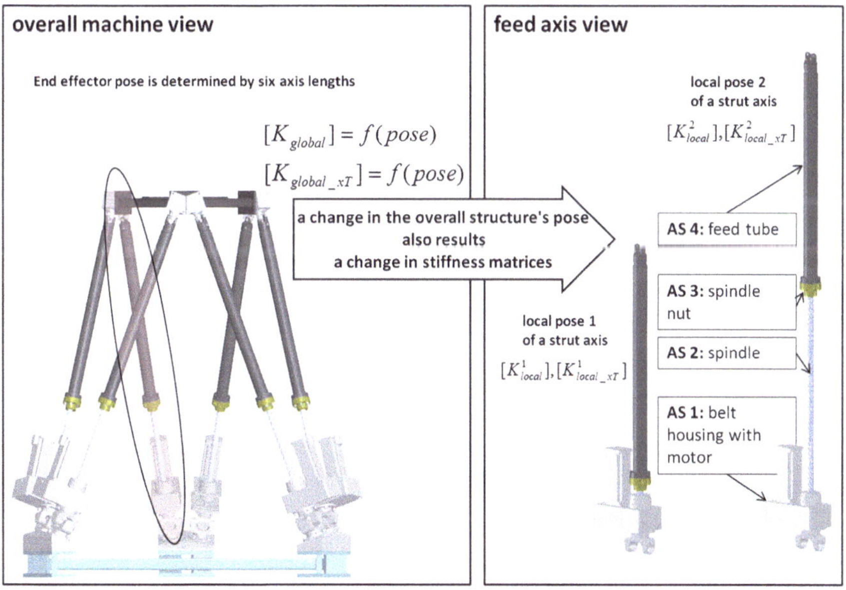

Fig. 7.2 Pose dependency of the mechanical and thermal stiffness matrices on a hexapod MiniHex (compare Chap. 1); detailed to local positions of feed axes. Subdivision of the feed axis into four stationary assemblies (*AS*)

$\{x^1\},\ldots,\{x^M\}$ are determined in real time by solving the simple system of linear Eq. (7.1b) only for these poses. The deformation in AS arrangements which are situated between the supporting positions is found via interpolation. This way full consistency with the correction model in subproject B07 (see Chap. 16) is achieved.

7.3.3.3 Thermo-Elastic Model

Distinguishing in the mechanical subproblem (7.1b) between subsystems that are independent of the temperature (7.10a) and those that are dependent on temperature (7.10b),

$$\{x\}_{static} = [K]^{-1}\{F\} \tag{7.10a}$$

$$\{x\}_{therm} = -[K]^{-1}[K_{xT}]\{T\} \tag{7.10b}$$

$$\{x\} = \{x\}_{static} + \{x\}_{therm} \tag{7.10c}$$

we find that (7.10b) has a structure identical to that of the second equation of the state space representation of the thermal subproblem (7.2) (with $[C]^T = -[K]^{-1}[K_{xT}]$). Thermal deformation $\{x\}_{therm}$ can thus be understood as an **additional output signal vector of the thermal problem**.

More exactly it means the following:

Let j be the fixed pose and $[K^j]$, $[K^j_{xT}]$ the related stiffness matrices. The complete MT structure consists of L assemblies so that the overall temperature vector $\{T\}$ is composed of temperature vectors of single AS$\{T\} = \{\{T^{BG_1}\},\ldots,\{T^{BG_L}\}\}$. Consequently, the thermal deformation in position j can be expressed as superposition as follows (compare (7.10b))

$$\{x^j\}_{therm} = \underbrace{\left[-[K^j]^{-1}[K^j_{xT}]\right]_{BG_1}}_{[C_x^{BG_1}]^T}\{T^{BG_1}\} + \cdots + \underbrace{\left[-[K^j]^{-1}[K^j_{xT}]\right]_{BG_L}}_{[C_x^{BG_L}]^T}\{T^{BG_L}\} \tag{7.11}$$

where $\left[-[K^j]^{-1}[K^j_{xT}]\right]_{BG_i} = [C_x^{BG_i}]^T$ are the elements of the complete matrix $[K^j]^{-1}[K^j_{xT}]$ the i-th AS temperature affects. Expanding the thermal AS-related model (7.2) using the equation $\{y_x^{BG_i}\} = [C_x^{BG_i}]^T\{T^{BG_i}\}$, the output signal vector $\{y_x^{BG_i}\}$ provides the partial deformation of the MT structure. The superposition of these signal vectors over all of the assemblies supplies then the total thermal deformation $\{x^j\}_{therm}$ in pose j.

The MOR method described above can be applied to the *AS-related thermo-elastic original system* Σ_x *extended this way*

$$\Sigma_x : \{\dot{T}\} + [A]\{T\} = [B]\{u\};\ \{y\} = [C]^T\{T\};\ \{y_x\} = [C_x]\{T\} \tag{7.12}$$

The extended reduced system will include the identical set of transformation matrices (compare (7.7)), because the Arnoldi procedure remains independent of the additional measurement matrix $[C_x]$ in the Krylov MOR.

Depending on requirements of each application, two approaches to deformation calculation can be distinguished: deformation can be obtained by solving the mechanical subtask (7.1b) or (7.10a, 7.10b, 7.10c) by using the complete temperature vector $\{T\}$ as input, whereby $\{T\}$ is first calculated by the solution of the reduced thermal subtask $\hat{\Sigma}$ and subsequent retransformation by means of Eq. (7.9). An alternative is to interpret thermal deformation as an additional output signal of the thermal system (as described above). The extended reduced system $\hat{\Sigma}_x$ provides the approximation of this additional output signal.

The two approaches are identical in terms of accuracy. However, the second variant is more efficient in terms of the computational efforts, since the time-consuming retransformation step (compare (7.9)) is unnecessary. This extended thermo-elastic MOR-FEM is particularly efficient if only the displacement at the TCP, rather than the total deformation, is of interest, as it is, for instance, in the case of the correction application (compare Chap. 16).

7.3.4 Practice Implementation of the Approach Shown for a Ball Screw Axis

This section elucidates the step-wise application of the methods described above developed for a typical machine tool structure of a ball screw axis of a hexapod MiniHex (compare Fig. 7.2). Starting from the CAD model, a rapid network model in form of Simulink block diagram (see Fig. 7.5) is developed.

Step 1 CAD preparation: First, the complete MT structure has to be subdivided into assemblies (AS) without relevant inner relative motions, see Fig. 7.2. Further, the procedure is continued in an AS-oriented manner according to the approach (Fig. 7.1).

The AS geometry can be simplified by removing thermo-elastically irrelevant geometry details.

Remark: Such irrelevant geometry details are typically removed before the FE analysis. The reason is to obtain a model with a justifiable quantity of degrees of freedom (DoF) during meshing with non-degenerated finite elements. This quantity is proportional to the calculation time for the (unreduced) FE model. However, during MOR-FE analysis, the reduced model is used for calculation. Thanks to the MOR procedure, the reduced model's DoF quantity is less by orders of magnitude.

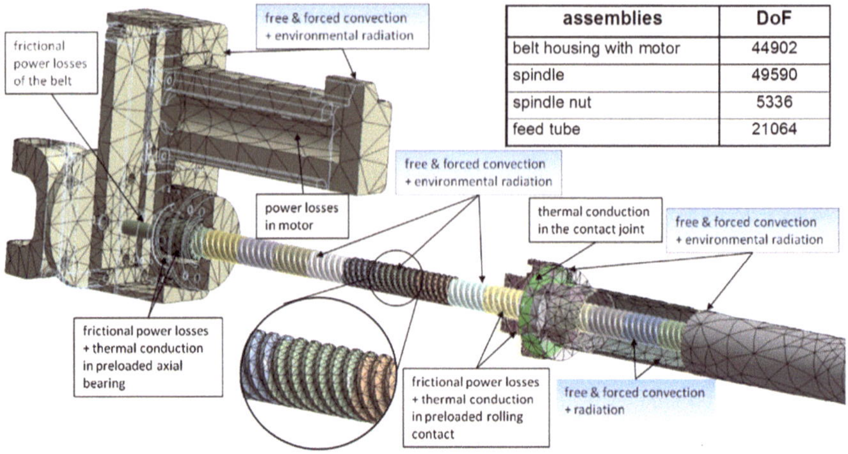

assemblies	DoF
belt housing with motor	44902
spindle	49590
spindle nut	5336
feed tube	21064

Fig. 7.3 AS-oriented thermal FE modelling with prepared segmented spindle surface to enter position-dependent heat flows

As a consequence, the mostly manual operation of geometry simplification no longer plays an important role, and can, under certain circumstances, even be omitted if appropriate automatic FE meshing tool is available.

Step 2 Thermal FE modelling: The AS geometry is imported into the FE software (for instance ANSYS), meshed, and the thermal material parameters (density, specific heat capacity and thermal conduction) are assigned. Then load types are applied to the AS—as a rule, on surfaces. Thereby convective loads and miscellaneous impressed heat flows are classified (compare Fig. 7.3).

A particular issue is the handling of surfaces through which heat flows are introduced depending on the relative AS position, such as spindle surfaces that are in frictional contact during the nut motion. These surfaces are cut into equidistant segments, see Fig. 7.3. One heat flow can be assigned to each segment. Thus it is possible to reconstruct the heat source motion, and thus it is completely consistent with the "local assignment" method in Chap. 16.

Ultimately, the matrices of the thermal FE model (7.2) are exported from ANSYS by means of in-house developed automatic ANSYS export routines (Großmann et al. 2011).

Step 3 Generation of Simulink-AS objects: A special library of "thermo-elastic network models" was developed and implemented (Fig. 7.4) for the block-oriented digital simulation (DBS) domain under Simulink. This library makes it possible to link the AS objects (block "AS") mapped in a reduced way into the thermo-elastic network model, thereby considering the structural variability (compare Fig. 7.5).

When creating the AS block, the FE model exported from ANSYS is imported to Matlab/Simulink by using in-house developed routines and transformed into the reduced model by means of the MOR methods mentioned above.

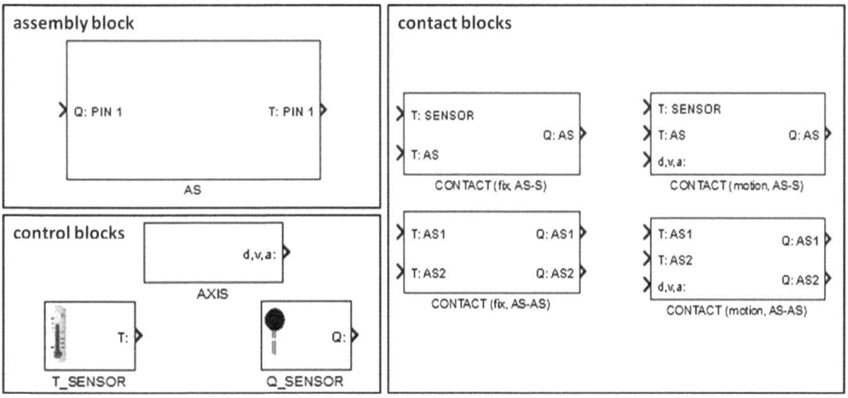

Fig. 7.4 Blocks of the Simulink library "thermo-elastic network models"

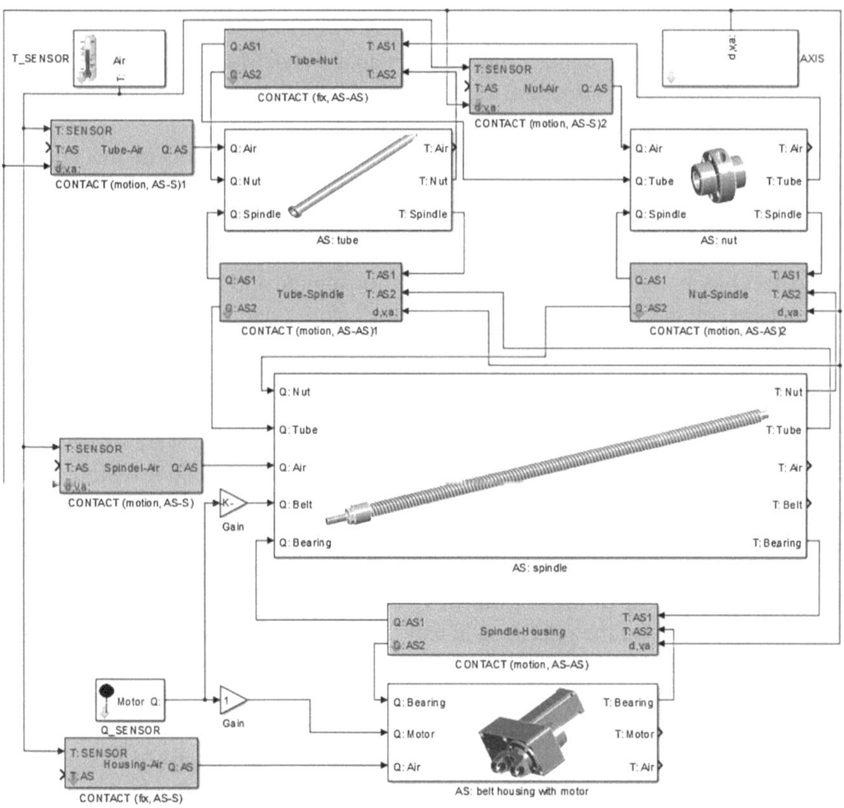

Fig. 7.5 Network model of the ball screw based strut axis mapped as SIMULINK block diagram

Step 4 Insertion of control blocks: Special control blocks were developed for the library of "thermo-elastic network models": a "temperature sensor" block makes it possible to define the ambient temperature (see Fig. 7.4); another sensor block, the "heat flow sensor", provides defined heat flows as outputs.

An axis block represents a feed axis object returning typical axis process parameters, such as the motor torques of the axis drive, as well as the position and velocity in the axis coordinate system (compare Fig. 7.4).

Step 5 Initial positioning: At first, each AS has a local coordinate system (CS). For each AS object, the location of the local CS in the global CS is defined in relation to a "zero-AS arrangement" of the machine tool. Here the zero-AS arrangement means just the pose, in which the relative mutual position of the moved assemblies (AS) is zero. This is coherent to a $\{0, 0, 0\}^T$ position in the machine coordinate system of an MT.

The positions of the sensor and axis objects in the global CS are specified analogously.

Step 6 Linking into a network model: The AS blocks are linked with each other and, if necessary, with the ambient (sensor) blocks by means of the contact blocks, see Fig. 7.5. The contact block represent a contact-load condition (compare Sect. 7.3.3 "Handling of structural variability"). Depending on the contact type, it is possible to choose among four contact blocks (compare Fig. 7.4).

Step 7 Parameter setting: The transient thermal calculation in the network model is carried out in the AS blocks at the level of reduced models. For this calculation, input signals, which are load parameters of the thermal model, are required. These load parameters are specified in the contact blocks. In the spindle-nut contact block, for instance, the heat flows for spindle and nut are each generated as functions of the axis parameters "position" and "velocity". Heat conduction between spindle and nut is also considered, and the boundary temperatures of the frictional contact partners enter the heat flow calculation (compare Fig. 7.5).

The thermal network model is ready for simulation after step 7 when the input data (kinematic regime, external heat flows, ambient temperatures, etc.) are supplied.

If the network model is only supposed to be run "off-line", then it is sufficient to supply input data captured by data logging from real or emulated operational scenarios.

For correction applications, however, the network model has to run "online" directly in the control of an MT. In this case, the input data are read out of the machine's CNC during the operation in real time, and are transmitted to the network model. This variant was demonstrated successfully for the network model of the hexapod MiniHex at the technical exhibition SPS-IPC-Drives 2013 (Großmann et al. 2013) and at the 2nd Vienna Production Technology Congress (Großmann et al. 2014b). Details about the control core—model interface are described in Chap. 16.

Step 8 Mechanic FE modelling: For the extended thermo-elastic AS network model, the extended measurement matrix $[C_x^{BG_i}]$ has to be generated according to Eq. (7.12) by means of the FE domain. For this purpose, the AS geometries meshed under step 2 are imported to ANSYS, where they are brought into the pose-specific

AS arrangement and the mechanical material data and load parameters are assigned. Afterwards, the pose-dependent stiffness matrices $[K^{Pose_j}]$, $[K_{xT}^{Pose_j}]$, as well as the static force vectors $\{F^{Pose_j}\}$ are exported from ANSYS analogously to step 2. This process is repeated for each pose. For example, for a ball screw, the corresponding matrices are generated for two local poses (compare Fig. 7.2, right).

Step 9 Extension of the AS objects: Analogously to step 3, the mechanical part of the extended thermo-elastic model is imported to Matlab/Simulink and transformed into the reduced form.

Thermo-elastic extension means that the AS block has an additional output$\{y_x\}$. This vector consists of triples of part deformation in x, y and z. The superposition of these part deformations provides the thermal deformation at the TCP in the poses considered.

7.3.5 Calculation Results and Performance

The results are demonstrated for two typical MT structures. The first is a column-spindle head structure described in Großmann et al. (2012a) and in Galant (2013), see also Fig. 7.1. The second example is the ball screw based strut axis of the hexapod MiniHex (see Fig. 7.2).

7.3.5.1 ANSYS Versus Matlab/Simulink Calculation

The variable loads due to the movement of the spindle head were simulated on the machine column.

The outcomes of a comparison between a conventional unreduced calculation using ANSYS and the reduced network model calculation are summarised in Fig. 7.6. A more detailed explanation is given in Großmann et al. (2012a, b). The transient simulated temperatures of the unreduced original FE model and the reduced network model are almost congruent; whereby the reduced variant provides savings in computing time of several orders of magnitude (18,000 s. vs. 18 s.).

7.3.5.2 Simulation Versus Experiment: Column-Spindle Head

For the same column-spindle head structure, loads were simulated on the machine column investigated on a test bed with 13 switchable heating films for each guiding rail. The variable loads again resulted from spindle head movement. The test bed is depicted in detail in Großmann et al. (2014a) and in Chap. 21.

The comparison of measurement versus simulation of the reduced network model indicates only slight deviations, see Fig. 7.7.

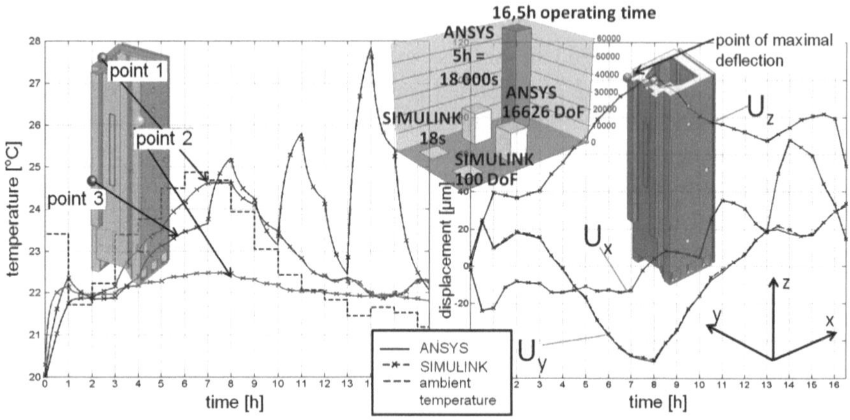

Fig. 7.6 Outcomes and performance of the unreduced and reduced models in comparison. Temperature (*left*) and displacement (*right*) in selected reference points

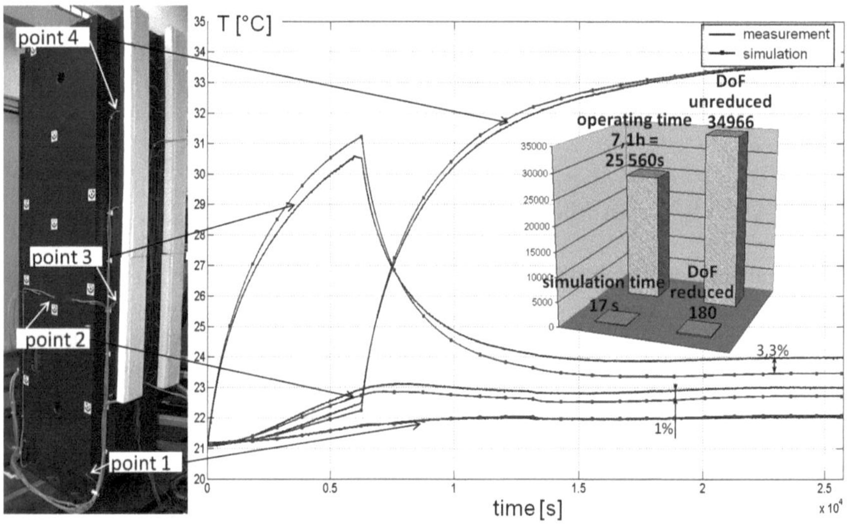

Fig. 7.7 Column-spindle head structure: measured versus simulated temperatures by using the high-speed reduced model

7.3.5.3 Simulation Versus Experiment: Ball Screw Based Strut Axis

The reduced network model of the MiniHex was immediately implemented into the machine control and validated by measurements in cooperation with the subprojects B04, B07, C04 and C06. The comparison results for simulation versus measurement demonstrated by the temperatures of two strut axes at the same kinematic regime are illustrated in Fig. 7.8 and coincide significantly.

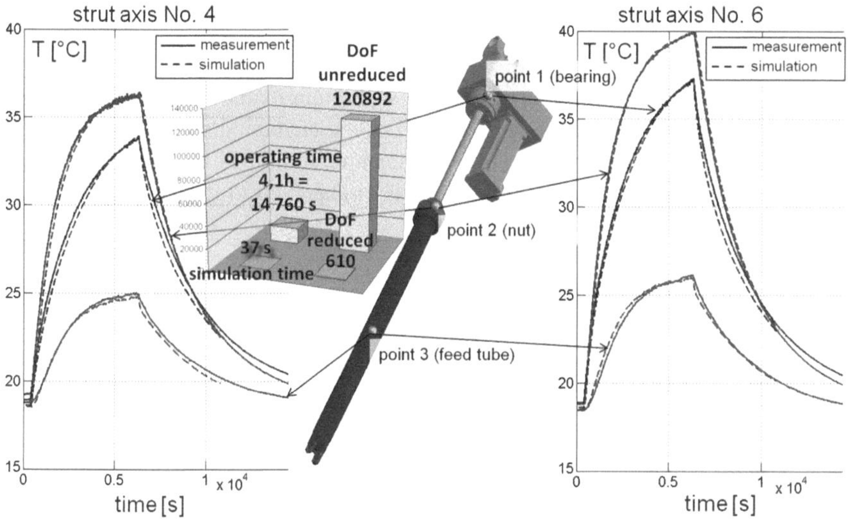

Fig. 7.8 Ball screw feed axis: measured temperatures versus temperatures obtained from simulations using the network model in real time

7.4 Classification Among the Objectives of the CRC/TR 96

The studies and research performed in the subproject A05 regarding the reduced thermo-elastic network model provide essential prerequisites for structure-variable modelling, particularly regarding the performance in terms of calculating time, permitting a structural model-based correction that explicitly represents the overall thermo-elastic functional chain to be implemented in the first place. For this reason, subproject A05 is closely aligned with subproject B07, which deals with structural model-based correction.

Expertise in parameter setting for power dissipations and heat transfers, mainly from the subprojects B04 and C04, is an additional contribution to the activities of subproject A05.

It is essential to validate the algorithms and models from A05 in experiment; this validation was conducted in cooperation with the work carried out in C06 and B07.

Ideas derived from subproject A06, especially regarding the parameter preserving model order reduction studied and developed in A06, were adopted and contributed to the network modelling algorithms.

7.5 Outlook

In the follow-up phases, the subproject A05 will retain its function as a lead project bringing together the partial models, as well as the technologies for modelling and parameter setting. Additional examples of machine tools will be explored, and the complexity of the associated network models is expected to increase significantly.

References

ANSYS, Inc. (2014) Help topics. Offline-Hilfe des Programmpakets ANSYS, Teil coupled effects

Arnoldi WE (1951) The principle of minimized iterations in solution of the matrix eigenvalue problem. Q Appl Math 9:17–29

Benner P, Mehrmann V, Sorensen DC (2005) Dimension reduction of large-scale systems. Lecture notes in computational science and engineering, vol 45. Springer, Heidelberg

Galant A (2013) Automatisierte Synthese blockorientierter Simulationsmodelle für die effiziente Berechnung thermo-elastischer Verformungen an Werkzeugmaschinen bei Berücksichtigung großer Relativbewegungen. In: Ansys conference & 31st CADFEM users' meeting 2013, Mannheim, 19–21 June 2013

Großmann K, Mühl A (2010) Reduktion strukturdynamischer und thermoelastischer FE-Modelle. ZWF 105(6):594–599

Großmann K, Galant A, Mühl A (2011) Model order reduction (MOR) for thermo-elastic models of frame structural components on machine tools. In: Ansys conference & 29th CADFEM users' meeting 2011, Stuttgart, 19–21 Oct 2011

Großmann K, Städel Ch, Galant A, Mühl A (2012a) Berechnung von Temperaturfeldern an Werkzeugmaschinen. Vergleichende Untersuchung alternativer Methoden zur Erzeugung kompakter Modelle. ZWF 107(6):452–456

Großmann K, Galant A, Mühl A (2012b) Effiziente Simulation durch Modellordnungsreduktion. Thermo-elastische Berechnung von Werkzeugmaschinen-Baugruppen. ZWF 107(6):457–461

Großmann K, Kauschinger B, Galant A, Thiem X et al (2013) Messestand MiniHex mit Präsentation der Teilprojekte A05, B04, B07, C06. Messe SPS-IPC-DRIVES, 26–28 Nov 2013

Großmann K, Galant A, Merx M, Riedel M (2014a) Verfahren zur effizienten Analyse des thermo-elastischen Verhaltens von Werkzeugmaschinen. In: 3rd International Chemnitz Manufacturing Colloquium ICMC 2014 "Innovations of Sustainable Production for Green Mobility", Reports from the IWU, vol 80, pp 683–700

Großmann K, Kauschinger B, Merx M, Riedel M, Galant A (2014b) Effiziente modell- und experimentgestützte Analyse des thermischen Verhaltens von Werkzeugmaschinen. 2.Wiener Produktionstechnik Kongress, 07–08 May 2014. Wien, Tagungsband, SS. 271–280

Groth C, Müller G (2009) FEM für Praktiker. Band 3: Temperaturfelder. Expert-Verlag

Krylov AN (1931) On the numerical solution of the equation by which in technical questions frequencies of small oscillations of material systems are determined. Izvestija AN SSSR (News of Academy of Sciences of the USSR), Otdel mat i estest nauk 7(4):491–539

Lohmann B, Salimbahrami B (2004) Ordnungsreduktion mittels Krylov-Unterraummethoden. at Automatisierungstechnik 52(1):30–38

Schilders W, van der Vorst H, Rommes J (2000) Model order reduction: theory, research aspects and applications. Springer, Heidelberg

Chapter 8
Model Order Reduction for Thermo-Elastic Assembly Group Models

Norman Lang, Jens Saak and Peter Benner

Abstract We present two model order reduction approaches based on different modelling strategies for a thermo-elastic assembly group model. Here, we consider the **machine stand example** given in Chap. 7. The focus is on capturing the structural variability. Therefore, we compare a switched linear systems (SLS) approach based on reduced order models determined by the Balanced Truncation (BT) method and a parametric model order reduction (PMOR) scheme based on an interpolatory projection method via the iterative rational Krylov algorithm (IRKA). In order to avoid the high dimensional coupled thermo-elastic system, additionally a Schur complement representation is applied to exploit the special structure of the one-sided coupling property of the system. The results show that both methods generate relative errors in the range of one per thousand.

8.1 Introduction

Throughout this contribution we consider the thermo-elastic model of the machine stand described in Chap. 7. The movement of the tool slide changes the contact boundary conditions of the heat transfer model. Therefore, we consider the equation

$$E_{th}\dot{T} = \mathcal{A}_{th}T + \mathcal{B}_{th}z \tag{8.1}$$

N. Lang (✉) · J. Saak · P. Benner
Mathematics in Industry and Technology, Department of Mathematics,
University of Technology, Chemnitz, Germany
e-mail: lang@mathematik.tu-chemnitz.de

J. Saak
e-mail: saak@mathematik.tu-chemnitz.de

P. Benner
e-mail: benner@mathematik.tu-chemnitz.de

K. Großmann (ed.), *Thermo-energetic Design of Machine Tools*,
Lecture Notes in Production Engineering, DOI 10.1007/978-3-319-12625-8_8

to model the temperature field $T \in \mathbb{R}^n$ of the machine stand. The operators $\mathcal{A}_{th} \in \mathbb{R}^{n\times n}$ and $\mathcal{B}_{th} \in \mathbb{R}^{n\times m}$ denote the variable system and input matrices of the thermal model for the machine stand, depending on the slide position, respectively. The tool slide is in full contact at every time instance t and therefore is described by a constant model with respect to the time. Therefore, it is sufficient to restrict ourselves to the heat model (8.1) of the machine stand for illustrating the structural variability. To this end, we treat the slide temperature implicitly via the input z. The input $z \in \mathbb{R}^m$ models the thermal loads influencing the system. In addition to the heat transfer model we consider

$$0 = A_{thel}T - A_{el}u \tag{8.2}$$

to model the stationary linear elastic behavior of the stand structure described by the deformation $u \in \mathbb{R}^{3n}$. Here $A_{el} \in \mathbb{R}^{3n\times 3n}$ denotes the elasticity system matrix and $A_{thel} \in \mathbb{R}^{3n\times n}$ is the coupling matrix which describes the thermally driven distortions in the elasticity model (8.2). Both matrices in Eq. (8.2) are associated to the same FE grid as the heat model (8.1). Therefore, the dimension of the elasticity model is $3n$ due to the x, y, z—directions of the deformation in every FE node. Note, there is no direct impact of the slide movement on the elasticity model and therefore, there is no change in the model with respect to the position of the tool slide. This leads to the coupled thermo-elastic system

$$\begin{aligned}\begin{bmatrix} E_{th} & 0 \\ 0 & 0 \end{bmatrix}\begin{bmatrix} \dot{T} \\ \dot{u} \end{bmatrix} &= \begin{bmatrix} \mathcal{A}_{th} & 0 \\ A_{thel} & -A_{el} \end{bmatrix}\begin{bmatrix} T \\ u \end{bmatrix} + \begin{bmatrix} \mathcal{B}_{th} \\ 0 \end{bmatrix} z \\ \Leftrightarrow: E\dot{x} &= \mathcal{A}x + \mathcal{B}z \end{aligned} \tag{8.3}$$

of dimension $4n$, with the state $x = \begin{bmatrix} T \\ u \end{bmatrix} \in \mathbb{R}^{4n}$. Furthermore, an output equation

$$y = \begin{bmatrix} 0, & C_{el} \end{bmatrix}\begin{bmatrix} T \\ u \end{bmatrix} = Cx \tag{8.4}$$

is needed to identify the degrees of freedom (DOF) that correspond to the locations of, e.g., measurement sensors. Since we are interested in the deformations of the machine stand, we only observe some DOFs corresponding to the deformations at the surface of interest. Therefore, $C_{el} \in \mathbb{R}^{q\times 3n}$ is the elasticity output matrix. The task now is to capture the structural variability by the reduction methods we use. Therefore, we consider parametric model reduction in Sect. 8.3 and an approach for switched linear systems in Sect. 8.2. The particular structure of the matrices $\mathcal{A}_{th}, \mathcal{B}_{th}$ will be shown in the corresponding sections.

For a fixed relative positioning of the machine stand and the tool slide, i.e., $\mathcal{A} := A, \mathcal{B} := B$ are time invariant, the associated system will lead to a generalised linear time invariant (LTI) state-space system of the form

$$\begin{aligned} E\dot{x} &= Ax + Bz, \\ y &= Cx \end{aligned}$$

with constant matrices $E, A \in \mathbb{R}^{4n\times 4n}, B \in \mathbb{R}^{4n\times m}$ and $C \in \mathbb{R}^{q\times 4n}$. The idea of projection based MOR now is to find projection matrices $V, W \in \mathbb{R}^{4n\times r}$, such that a reduced order model

$$\begin{aligned} \hat{E}\dot{\hat{x}} &= \hat{A}\hat{x} + \hat{B}z, \\ \hat{y} &= \hat{C}\hat{x} \end{aligned}$$

of dimension $r \ll 4n$ with $\hat{y} \approx y$ can be achieved. The reduced order matrices are computed in the form

$$\begin{aligned} \hat{E} &:= W^T E V \in \mathbb{R}^{r\times r}, \quad \hat{A} := W^T A V \in \mathbb{R}^{r\times r}, \\ \hat{B} &:= W^T B \in \mathbb{R}^{r\times m}, \quad \hat{C} := C V \in \mathbb{R}^{q\times r}. \end{aligned}$$

For the more advanced task of parameter dependent matrices $E(\mu), A(\mu), B(\mu), C(\mu)$ we consider systems of the form

$$\begin{aligned} E(\mu)\dot{x} &= A(\mu)x + B(\mu)z, \\ y &= C(\mu)x, \end{aligned} \tag{8.5}$$

where the quantity μ describes the parameter that we want to preserve within the model order reduction process. Therefore, without loss of generality (see e.g., Haasdonk and Ohlberger 2009), we consider the parameter dependence in the system matrices to be of an affine form

$$E(\mu) = E_0 + e_1(\mu)E_1 + \cdots + e_{M_E}(\mu)E_{M_E}, \quad M_E \in \mathbb{N}. \tag{8.6}$$

Analogously, the matrices $A(\mu), B(\mu), C(\mu)$ are represented by parameter dependent functions $f_j(\mu), g_j(\mu), h_j(\mu)$ and the numbers $M_A, M_B, M_C \in \mathbb{N}$, respectively. M_E, M_A, M_B, M_C denote the numbers of affine summands to define E, A, B, C. Exploiting the parameter affine structure of the matrices allows us to compute reduced matrices

$$\hat{E}_j := W^T E_j V, \quad j = 1, \ldots, M_E, \tag{8.7}$$

while the parameter evaluation can be performed on the reduced order level afterwards by evaluating the functions $e_j(\mu), j = 1, \ldots, M_E$. Analogously, the matrices $A(\mu), B(\mu), C(\mu)$ can be reduced independent of μ and evaluated with respect to the parameter by f_j, g_j, h_j, respectively. Note that the numbers M_A, M_B, M_C of summands in the analogues for $A(\mu), B(\mu), C(\mu)$ of Eq. (8.6) do not necessarily need to be equal.

In order to avoid computations based on the coupled thermo-elastic model of dimension $4n$ we want to exploit the special structure given in (8.3), i.e., we make use of the one-sided coupling. Since we assume no coupling from the elasticity model in the direction of the thermal model, we have a zero block of size $n \times 3n$ in the upper right corner of the system matrix A of Eq. (8.3). Applying the Schur complement to Eqs. (8.3) and (8.4) for fixed $\mathcal{A}_{th} := A_{th}$ and $\mathcal{B}_{th} := B_{th}$ as proposed in Freitas et al. (2008) in advance of the MOR procedure we obtain the system

$$\begin{aligned} E_{th}\dot{T} &= A_{th}T + B_{th}z, \\ y &= C_{el}A_{el}^{-1}A_{thel}T = \bar{C}T. \end{aligned} \tag{8.8}$$

Now the system (8.8), for which we want to find a reduced order approximant, is again of order n which is the dimension of the heat transfer model. That is, due to the special structure of the coupled thermo-elastic model, here we need to pay particular attention to the modified output matrix $\bar{C}$. The matrices V and W, we are looking for now, have dimension $n \times r$.

8.2 MOR for Switched and Coupled Systems

Since the actual position of the slide is described by a time dependent trajectory, one can think of a special case of parameter dependence, namely, time variability. Here linear time varying (LTV) systems become the focus of attention. Model reduction for LTV systems is a highly storage consuming process. Therefore, here we use detailed information given by the model described in Chap. 7. Since the guide rails are modelled by 15 segments and additionally the slide follows discrete positions during the movement, we set up a switched linear system (SLS). Due to the specific height of the tool slide and the fixed mesh resolution on the guide rails, the contact boundary can only change within a fixed number of possible setups. In particular we know that at every time step, the slide covers exactly 5 segments on the rail. This in fact allows 11 discrete positions which can be reached and we consider the heat transfer model to be a SLS of the form

$$E_{th}\dot{T} = A_{th}^{\alpha}T + B_{th}z^{\alpha}. \tag{8.9}$$

Here, α is a piecewise constant function of time, which takes its value from the index set $\mathbf{J} = \{1, \ldots, 11\}$. The function $\alpha(t)$ denotes a switching signal, representing the active boundary part at time t. In other words, each discrete stand-slide-configuration is determined by α. Together with Eqs. (8.2) and (8.4) and applying the Schur complement representation (8.8) we obtain the system

$$
\begin{aligned}
E_{th}\dot{T} &= A_{th}^{\alpha}T + B_{th}z^{\alpha}, \\
y &= \bar{C}T.
\end{aligned}
\tag{8.10}
$$

Here, the change of the input operator $\mathcal{B}_{th} := B_{th}$ is hidden in the input z^{α} itself. In order to obtain a reduced order model approximating (8.8), we compute sets of projection matrices V_j, W_j for each fixed position $j = 1, \ldots, 11$ of the tool slide using the Balanced Truncation (BT) method (for details on BT see e.g., Enns 1984; Laub et al. 1987; Moore 1981). Therefore, every stand-slide setup is modelled by one of these subsystems. Finally, we need to decide which subsystem needs to be used during the online simulation via the switching signal $\alpha(t)$. For a deeper insight into BT model reduction for switched systems we refer the reader to e.g., Geuss and Diepold (2013), Monshizadeh et al. (2011), and the references therein.

For verification of the proposed methods, we follow the simulation example given in Chap. 7. That is, we solve the dynamic heat model (8.9) on the time interval $[0, 16.5]$ (h) with a fixed step size of $\tau = 10$ (s). The full order model is of dimension $n = 16{,}626$ and we also consider the outputs described in Chap. 7. This in fact means that we observe 27 deformation DOFs via $C_{el} \in \mathbb{R}^{27\times 3n}$. Here, the trajectory of the full order system is compared to the trajectories produced by reduced order systems of size $10, 40$ and 60 [see Fig. 8.1 (left)] in each possible slide position. In the comparison we consider Node 1 of the description in Chap. 7. To quantify the results the relative errors are presented in Fig. 8.1 (right). For a reduced order $r = 60$ a relative error in the range of one per thousand could be achieved, which is quite satisfactory. Note, the peak at hour 11 in the relative error plot is a purely numerical issue in the evaluation of the relative simulation error and is not problematic for practical applications.

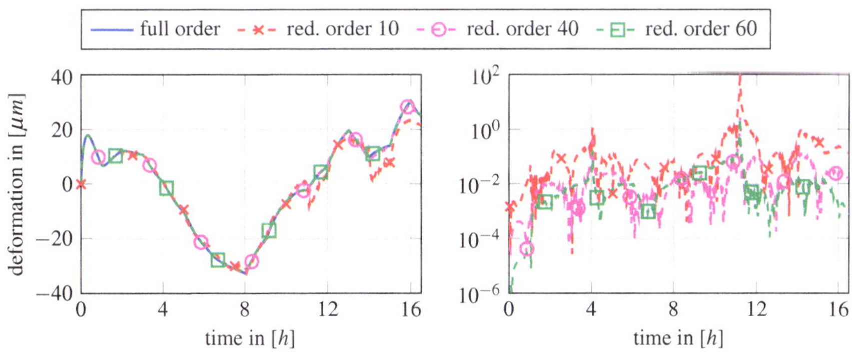

Fig. 8.1 Deformation progress and relative error of the full order system compared to reduced systems of order $r \in \{10,\ 40,\ 60\}$ represented in Node 1

8.3 PMOR by the Iterative Rational Krylov Algorithm

Since in case of the switched systems approach we need to compute a single reduced order system for each possible discrete position of the tool slide, which easily becomes highly memory demanding when more than just a few subsystems have to be considered, in this section we consider the varying position of the tool slide to be a parameter μ of the system. Depending on μ one has to decide which parts of the boundary are affected by the contact of the tool slide and the stand. Therefore, Eq. (8.1) becomes the parametric thermal state equation

$$E_{th}\dot{T} = A_{th}(\mu)T + B_{th}(\mu)z$$

of dimension n, with $\mathcal{A}_{th} := A_{th}(\mu)$ and $\mathcal{B}_{th} := B_{th}(\mu)$. Similar to the SLS (8.10), we use the Schur complement representation to exploit the one-sided coupling structure of the system (8.3) and we end up with the modified parametric model

$$\begin{aligned} E_{th}\dot{T} &= A_{th}(\mu)T + B_{th}(\mu)z, \\ y &= \bar{C}T. \end{aligned} \tag{8.11}$$

The remainder is a straightforward implementation of known theory for model order reduction by Krylov-based interpolatory projection methods shown in Baur et al. (2011). This means that for the computation of the required matrices V, W in this section we use the iterative rational Krylov algorithm (IRKA) following the theory given there. The main idea we adopt here is to simply use interpolatory model reduction for fixed parameter sample values $\mu_1, \ldots, \mu_k$, i.e., we generate pairs $V_j, W_j \in \mathbb{R}^{n\times r_j}, j = 1, \ldots, k$, which build $\mathcal{H}_2$ optimal reduced order models of dimension r_j in the parameter sample points (for details on $\mathcal{H}_2$ model reduction see e.g., Bunse-Gerstner et al. 2010; Gallivan et al. 2008; Gugercin et al. 2008). The local reduction bases V_j, W_j with respect to the parameter μ_j form the global reduction bases

$$V = [V_1, \ldots, V_k], \qquad W = [W_1, \ldots, W_k],$$

by concatenation. The global matrices $V, W \in \mathbb{R}^{n\times r}$ with $r = \sum_{j=1}^{k} r_j$ are used to compute the reduced order matrices $\hat{E}_j, \hat{A}_j, \hat{B}_j, \hat{C}_j$ in the parameter affine representation, feasible over the whole parameter range as shown in Eq. (8.7). Again, the reduced order model is set up by evaluating the parameter dependent functions e, f, g, h in Eq. (8.6) in the form

$$\hat{A}(\mu) = \sum_{j=1}^{M_A} f_j(\mu)\hat{A}_j = \sum_{j=1}^{M_A} f_j(\mu)W^T A_j V,$$

where M_A is the number of parameter affine summands for $\hat{A}(\mu)$. The matrix $\hat{B}(\mu)$ is computed analogously with $g_j(\mu)$ and M_B. Note that the Petrov-Galerkin projection by V, W produced by IRKA does not ensure stability of the reduced system. We use one-sided Ritz-Galerkin projection, i.e., $V = W$, since this is known to preserve the stability of the original system by Sylvester's law of inertia (see e.g., Golub and Van Loan 1996). For several numerical examples, in particular the one under investigation here, it is a common observation that this is indeed necessary.

As in the previous section, the progression of the deformation of the output nodes described in Chap. 7 is shown. Here, we have to determine the position of the slide at a certain time t to evaluate the parameter dependent functions in the affine representation (8.6) of the system (8.11). The following figures show the full order simulation of the thermo-elastic model compared to the reduced order trajectories [see Fig. 8.2 (left)] in Nodes 1–3, as well as the relative error of the deformation trajectories [see Fig. 8.2 (right)]. The orders $r_1, \ldots, r_k$ of the reduced systems in the chosen parameter sample points are equally chosen to be 25. Therefore, the order of the global reduced system with respect to the parameter range is kr_j. Here, we chose the $k = 3$ sample points as equally distributed over the parameter range, i.e., we get a reduced order of $r = 75$. The relative error plot shows that already for 3 sample points, the relative errors lie on average below the percent range. Similar to the results for the switched system approach the peaks here can be explained by the division by a number close to zero in the relative error.

Comparing the results in Sects. 8.2 and 8.3 we observe that the SLS approach solves the problem in less time compared to the PMOR method for both, the full order and reduced order system, respectively. This is due to the fact that for the SLS scheme we have exactly 11 possible stand-slide configurations and additionally we solve the heat model for a fixed time-step size. That means, the operator of the linear system, with which we have to solve in every time step, does not change and we can recycle the LU-decompositions of the linear system. In the case of the

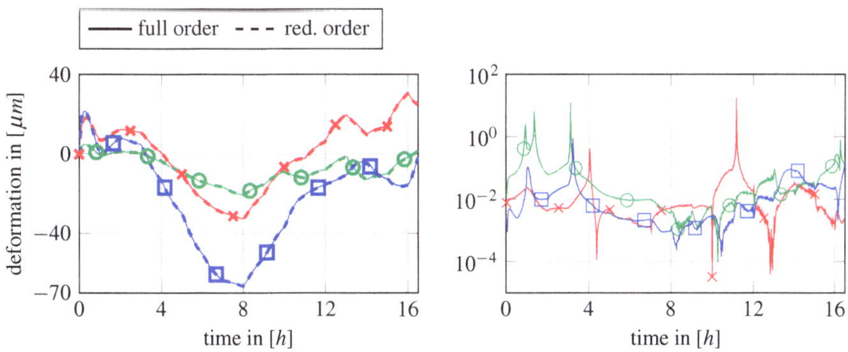

Fig. 8.2 Deformation progress and relative error in Nodes 1 (×), 2 (○) and 3 (□), described in Chap. 7, of the full order system compared to global reduced system with respect to the parameter μ of order $r = 75$

Table 8.1 Comparison of system orders, times for the reduction processes and the corresponding simulation times for the full and the reduced order systems for the SLS approach and the PMOR scheme, respectively

	System order	Time for reduction	Simulation time (s)
Full order SLS	16626	–	43.11
Reduced SLS	60	4560.32	0.69
Full order PAR	16626	–	1028.22
Reduced PAR	75	2090.56	1.76

parametric model the variability described by the parameter μ is considered to be continuous and therefore we do not have a fixed number of discrete positions as in the SLS case. That means, since the operator of the linear system changes continuously, we cannot recycle the LU-decompositions for solving the problem in a comparable simple way. On the other, choosing the most storage efficient method with respect to e.g., memory constraints on micro controllers directly depends on the number of subsystems and additionally the corresponding number of state-space transformations in the SLS method or the number of summands M_E, M_A, M_B, M_C in the parameter affine representation of the PMOR approach, respectively. Nevertheless, both methods lead to a drastic reduction of the simulation time compared to their full order simulations. A comparison of the system orders, the execution times for the reduction procedures and the corresponding simulation times for the full and reduced order models can be found in Table 8.1. The simulation approximates the heating process described in Chap. 7 over 16.5 (h) induced by the moving tool slide.

Regarding the computation times of the reduction procedures one should note that this is done in the offline phase, i.e., the projection matrices V, W are only computed once in advance of the online simulation. Furthermore, the reduction times can be further reduced using parallel algorithms. That is, since the reduced systems of all stand-slide configurations in the SLS approach can be computed completely independent, as well as the 3 reduced order models for the parameter sample points in the parametric ansatz, we can reduce the execution time by a factor of approximately 11 and 3, respectively.

8.4 Integration into the CRC/Transregio

Structural variability is an omnipresent problem in the CRC due to relatively moving assembly groups. In the context of heat transfer this immediately induces a moving load problem. Proper MOR for this type of task is an active research topic. We have investigated two approaches that allow us to find a moderate reduced order approximant of such kind of variable models. The first approach treats the special discrete structure of the model arising in Chap. 7. The second method rewrites the actual time dependence as a continuous parameter dependence. Both approaches enable the application of well-approved methods for linear time invariant systems. The explicit treatment of the actual linear time varying problem, although theoretically well understood for decades, requires additional research on the numerical methods. The main challenges for future activities are to investigate proper storage schemes and to find solvers for the underlying equations.

References

Baur U, Beattie CA, Benner P, Gugercin S (2011) Interpolatory projection methods for parameterized model reduction. SIAM J Sci Comput 33(5):2489–2518

Bunse-Gerstner A, Kubalinska D, Vossen G, Wilczek D (2010) h_2-norm optimal model reduction for large scale discrete dynamical MIMO systems. J Comput Appl Math 233(5):1205–1216

Enns DF (1984) Model reduction with balanced realizations: an error bound and a frequency weighted generalization. In: The 23rd IEEE conference on decision and control, vol 23, pp 127–132

Freitas F, Rommes J, Martins N (2008) Gramian-based reduction method applied to large sparse power system descriptor models. IEEE Trans Power Syst 23(3):1258–1270

Gallivan K, Vandendorpe A, Van Dooren P (2008) $\mathcal{H}_2$-optimal model reduction of MIMO systems. Appl Math Lett 21:1267–1273

Geuss M, Diepold KJ (2013) An approach for stability-preserving model order reduction for switched linear systems based on individual subspaces. In: Roppenecker G, Lohmann B (eds) Methoden und Anwendungen der Regelungstechnik. Shaker Verlag, Aachen

Golub G, Van Loan C (1996) Matrix computations, 3rd edn. Johns Hopkins University Press, Baltimore

Gugercin S, Antoulas AC, Beattie C (2008) $\mathcal{H}_2$ model reduction for large-scale dynamical systems. SIAM J Matrix Anal Appl 30(2):609–638

Haasdonk B, Ohlberger M (2009) Efficient reduced models for parametrized dynamical systems by offline/online decomposition. In: Proceedings of the MATHMOD 2009, 6th Vienna international conference on mathematical modelling

Laub AJ, Heath MT, Paige CC, Ward RC (1987) Computation of system balancing transformations and other applications of simultaneous diagonalization algorithms. IEEE Trans Autom Control 32(2):115–122

Monshizadeh N, Trentelman HL, Çamlibel MK (2011) Simultaneous balancing and model reduction of switched linear systems. In: The 50th IEEE conference on decision and control and European control conference, pp 6552–6557

Moore BC (1981) Principal component analysis in linear systems: controllability, observability, and model reduction. IEEE Trans Autom Control AC-26(1):17–32

Chapter 9
High-Accuracy Thermo-Elastic Simulation on Massively Parallel Computer

Andreas Naumann, Florian Stenger, Axel Voigt and Jörg Wensch

Abstract Subproject A07 develops and explores numerical methods and techniques to implement them to solve problems in connection with thermo-elastic subassemblies' and machine tools' simulation in the CRC/Transregio 96. For this purpose, high-resolution discretization methods were developed, tested and applied both for high resolution discretization in space and for efficient integration in the long term.

9.1 Introduction

The goal of this subproject entails the development of efficient methods to simulate the thermo-elastic performance of machine tools or their subassemblies under real load conditions. Modelling and parameter setting approaches covering the entire system of the machine tool and used for the validation of several techniques for model reduction are developed. A milling machine column and a spindle are chosen as exemplary subassemblies (see Fig. 9.1, Chap. 1). The geometry of both parts is represented by a CAD model. The spindle moves up and down at a given speed. Frictional heat emitted along the line of contact makes the machine tool designers interested in understanding the heat distribution within the machine tool and the resulting deformation.

A. Naumann (✉) · F. Stenger · A. Voigt · J. Wensch
Institut für Wissenschaftliches Rechnen, TU Dresden, 01062 Dresden, Germany
e-mail: andreas.naumann@tu-dresden.de

F. Stenger
e-mail: florian.stenger@tu-dresden.de

A. Voigt
e-mail: axel.voigt@tu-dresden.de

J. Wensch
e-mail: joerg.wensch@tu-dresden.de

K. Großmann (ed.), *Thermo-energetic Design of Machine Tools*,
Lecture Notes in Production Engineering, DOI 10.1007/978-3-319-12625-8_9

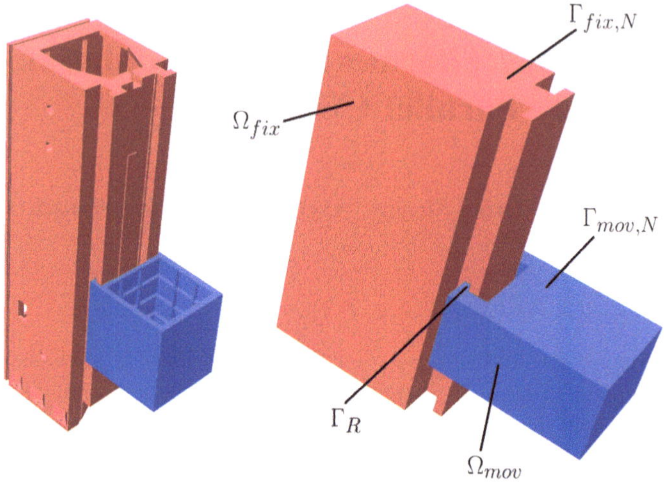

Fig. 9.1 The tool's original geometry (*left*) and the simplified geometry used for different simulation approaches (*right*). A spindle (*blue*, Ω_{mov}) and a column of a milling machine (*red*, Ω_{fix}) is shown, the spindle moves up and down the column. Neumann boundaries $\Gamma_{\cdot,N}$ are denoted and the contact line $\Gamma_R = \overline{\Omega}_{fix} \bigcap \overline{\Omega}_{mov}$

Two mathematical approaches are taken into account for the phenomena described above:

- Two different geometries, which are given independently, are sliding against each other during the process. This leads to a contact problem with varying contact conditions, which have to be considered numerically. Two different approaches will be compared. A diffuse domain approach, introduced in Li et al. (2009), in which the geometries and therefor also the contact area, are only implicitly described by a phase-field function on one global discretization, and a direct approach, in which the contact area is explicitly computed in each time step and an interpolation between two different meshes is used to deal with the heat exchange along the contact area. The different method and their comparison on a simplified geometry are described in Sect. 9.2.2.
- In the reference problem, the thermo-elastic machine tool behaviour was simulated over a period of 15 h. Here, the spindle head's movement occurs periodically in one-hour time increments, with periods from 1.5 to 48 s. The periods and velocities are shown in Table. 9.1. The timescale of the periods is clearly shorter than that of the integration. Section 9.2.3 deals with an approach that makes use of this periodicity in temporal integration. For numerical simulation, defect corrected averaging employs a replacement problem identified as suitable, which precisely approximates the solution in stroboscopic spaces.

Table 9.1 Velocities and corresponding periods

h	1	2	3	4	5	6	7	8	9	10	11	12	13	14	15
v (m/min)	2.5	2.5	2.5	5	5	5	10	10	10	15	15	15	20	20	20
Period (s)	12	48	12	6	24	6	3	12	3	2	8	2	1.5	6	1.5

9.2 Approaches

9.2.1 Mathematical Model for Heat Exchange

On the two domains Ω_{fix} and Ω_{mov}, where the first one is fixed and the second one represents the moving part, two temperatures T_{fix} and T_{mov} have to be considered. The equations read:

$$\rho c_p \partial_t T_{fix} - k\Delta T_{fix} = 0 \quad \text{in} \quad \Omega_{fix} \tag{9.1}$$

$$\rho c_p (\partial_t T_{mov} + \nabla \cdot (\mathbf{v} T_{mov}) - b\Delta T - mov = 0 \quad \text{in} \quad \Omega_{mov}(t) \tag{9.2}$$

where ρ is the materials density, c_p the specific heat capacity and k the thermal conductivity. $\mathbf{v}$ is the given velocity of the moving part. The boundary conditions

$$\nabla T_{fix} \cdot \mathbf{n} = 0 \quad \text{on } \Gamma_{fix,N}(t) \tag{9.3}$$

$$\nabla T_{fix} \cdot \mathbf{n} = \alpha(T_{fix} - T_{mov}) + \frac{\eta}{2} \qquad \text{on } \Gamma_R(t) \tag{9.4}$$

$$\nabla T_{mov} \cdot \mathbf{n} = 0 \quad \text{on } \Gamma_{mov,N}(t) \tag{9.5}$$

$$\nabla T_{mov} \cdot \mathbf{n} = \alpha(T_{mov} - T_{fix}) + \frac{\eta}{2} \quad \text{on } \Gamma_R(t), \tag{9.6}$$

with α the heat exchange parameter and η a source term due to friction, are applied. Both equations are coupled through the boundary condition on the moving contact line $\Gamma_R(t)$. On the remaining boundaries $\Gamma_{fix,N}$ and $\Gamma_{mov,N}$ a no-flux condition with $\mathbf{n}$ denoting the normal is assumed. Finally $\Omega_{mov}(0) = \Omega_{mov,0}$ and $T_{fix} = T_{mov} = T_0$ for the moving part and the temperatures are used as initial conditions respectively.

9.2.2 Spatial Discretization for Contact Problem in a Simplified Geometry

9.2.2.1 The Diffuse-Domain Method

A phase-field representation of the geometry is constructed in each time step and an implicit mesh defined in the larger box-domain Ω^{dd}, with an adapted grid resolution according to the position of the spindle. Three different were considered phase-field variables, one for the column ϕ_{fix}, one for the spindle $\phi_{mov}(t)$, and one for the

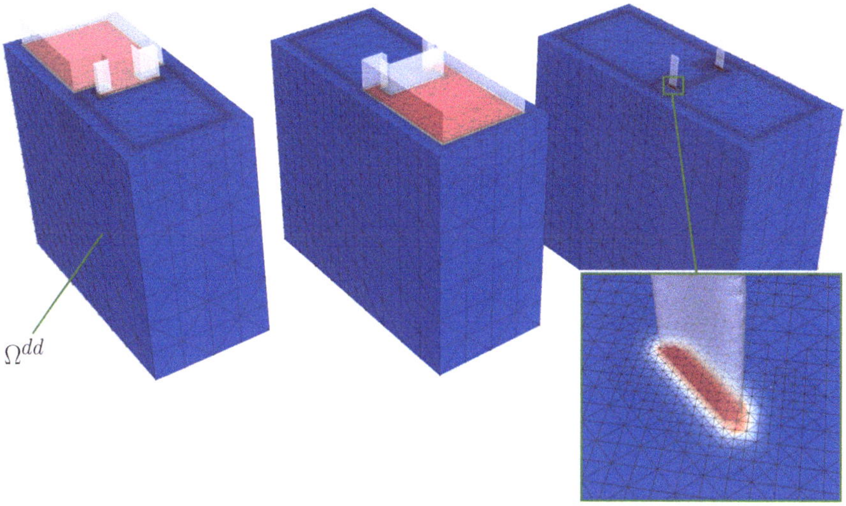

Fig. 9.2 Three clips through the implicit mesh showing the phase-fields for the column (*left*), the spindle (*middle*) and the interface between both parts (*right*). In each of the three pictures the respective component is shown by a transparent *grey* 0,5-contour. The mesh has approximately 1,2 Mio elements, the smallest edge is approximately 1.7 cm. Due to refinement and coarsening the number of elements remains almost constant during the simulation

boundary, where both domains are in contact $\phi_{bnd}(t)$. ϕ_{fix} and $\phi_{mov}(t)$ are defined according to

$$\phi(\mathbf{x},t) := \frac{1}{2}\left(1 - \tanh\left(\frac{3r(\mathbf{x},t)}{\varepsilon}\right)\right), \tag{9.7}$$

where $r(\mathbf{x},t)$ is the signed distance function to $\partial\Omega_{fix}$ or $\partial\Omega_{mov}(t)$. ε is the length scale determining the width of the diffuse interface. $\phi_{bnd}(t)$ results as the product of ϕ_{fix} and $\phi_{mov}(t)$. Figure 9.2 shows the three phase-field functions on distinct clips of the implicit mesh in $\Omega^{dd} = [0,400] \times [0,900] \times [0,900]$ in cm.

All quantities are extended to Ω^{dd} and the diffuse domain approximation is specified as

$$\rho c_p \phi_{fix} \partial_t T_{fix} - k\nabla \cdot (\phi_{fix} \nabla T_{fix}) + \frac{1}{\varepsilon} B(\phi_{bnd})(\alpha(T_{fix} - T_{mov}) + \frac{\eta}{2}) = 0 \tag{9.8}$$

$$\rho c_p(\partial_t(\phi_{mov} T_{mov}) + \nabla \cdot (\phi_{mov} \mathbf{v} T_{mov})) - k\nabla \cdot (\phi_{mov} \nabla T_{mov}) \\ + \frac{1}{\varepsilon} B(\phi_{bnd})(\alpha(T_{mov} - T_{fix}) + \frac{\eta}{2}) = 0 \tag{9.9}$$

in $\Omega^{dd} \cdot \varepsilon^{-1} B(\phi_{bnd}) = \varepsilon^{-1} \phi_{bnd}^2 (1 - \phi_{bnd})^2$ serves as an approximation for the delta-function of Γ_R. The initial conditions are specified as $\nabla T_{fix} \cdot \mathbf{n} = \nabla T_{mov} \cdot \mathbf{n} = 0$ on $\partial\Omega^{dd}$ and $T_{fix} = T_{mov} = T_0$.

9.2.2.2 Explicit Contact Formulation

The two coupled heat transfer equations (9.1) and (9.2) with their corresponding boundary conditions (9.3) to (9.6) are discretized independently of each other by means of linear finite elements in space. They appear in the corresponding weak formulation

$$\begin{aligned} T_{fix}(x,t) &= \sum_i T_{fix,i}(t)\varphi_{fix,i}(x) \\ T_{mov}(x(t),t) &= \sum_i T_{mov,i}(t)\varphi_{mov,i}(x(t),t) \\ \forall j : \rho c_p\left(\partial_t T_{fix}, \varphi_{fix,j}\right) + k\left(\nabla T_{fix}, \nabla \varphi_{fix,j}\right) &= \left(\frac{\eta}{2} + \alpha\left(T_{fix} - T_{mov}\right), \varphi_{fix,j}\right)_{\Gamma_R(t)} \\ \forall j : \rho c_p\left(\partial_t T_{mov}, \varphi_{mov,j}\right) + k\left(\nabla T_{mov}, \nabla \varphi_{mov,j}\right) &= \left(\frac{\eta}{2} + \alpha\left(T_{mov} - T_{fix}\right), \varphi_{mov,j}\right)_{\Gamma_R(t)} \end{aligned}$$

with the test functions φ_{mov} and φ_{fix} as well as the commonly used L^2-scalar product $(f,g) = \int_\Omega f\, g dx$ with $\Omega = \Omega_{fix}, \Omega_{mov}$. The difficulty arises from the approximation of the boundary integrals and the coupling of T_{mov} and T_{fix} at the contact area.

The time dependency of the boundary integration area is represented by means of the time dependent indicator function

$$S(t,x) = \begin{cases} 1 & x \in \Gamma_R(t) \\ 0 & \text{else} \end{cases} .$$

Thus the boundary integrals are extended to the whole frictional boundary $\Gamma_R = \bigcup_{t \in [0,\infty]} \Gamma_R(t)$, independent of the time.

The corresponding boundary integrals

$$\begin{aligned} \left(\frac{\eta}{2} + \alpha(T_{fix} - T_{mov}), \varphi_{fix,j}\right)_{\Gamma_R(t)} &= \left(S(t,x)\left(\frac{\eta}{2} + \alpha(T_{fix} - T_{mov})\right), \varphi_{fix,j}\right)_{\Gamma_R} \\ &= (F(t,x), \varphi_{fix,j})_{\Gamma_R} \end{aligned}$$

now only contain the time dependent non-linear function $F(t,x)$. This function includes the temperature difference $T_{mov} - T_{fix}$, whose evaluation is illustrated in Fig. 9.3 for the example of $T_{fix}(c_i)$ on the moved boundary of Ω_{mov}. The value of

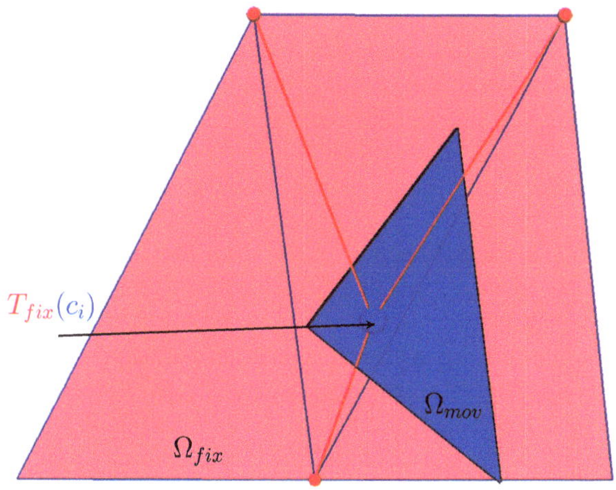

Fig. 9.3 Interpolation of the solutions fo Ω_{fix} (*red*) in the quadrature point c_i of the moved grid Ω_{mov} (*blue*) at the contact area

T_{fix} at the quadrature point c_i is determined by the interpolation of the basis functions on Ω_{fix}.

Spatial discretization followed by the special processing of the contact boundary condition produces an ordinary differential equation of the type

$$\begin{pmatrix} M_{fix} & 0 \\ 0 & M_{mov} \end{pmatrix} \begin{pmatrix} T_{fix} \\ T_{mov} \end{pmatrix} = \begin{pmatrix} A_{11}(t) \, A_{12}(t) \\ A_{21}(t) \, A_{22}(t) \end{pmatrix} \begin{pmatrix} T_{fix} \\ T_{mov} \end{pmatrix} + \begin{pmatrix} G_1(t) \\ G_2(t) \end{pmatrix} \tag{9.10}$$

including the mass matrices M_{fix} and M_{mov}, the time dependent stiffness matrices A_{ij} and the friction terms $G_j(t)$. The resulting system of ordinary differential equations is solved by suitable time integration techniques. Because this is a stiff problem, implicit techniques have to be employed. The Rosenbrock methods have particularly proved to be the method of choice.

9.2.3 *Efficient Long-Term Integration of the Column Geometry*

9.2.3.1 Defect Corrected Averaging

When interpreting the thermal or thermo-elastic machine tool simulation according to the method of lines, then, after having performed the spatial discretization, an ordinary differential equation (9.10) is obtained. A notable feature in the case of the spatially discretized machine tool is a periodic source term on the right side, whereas the period is very short versus temporal horizon of simulation. It is costly and labour-intensive to subject the problem to long-term integration with simultaneous resolution of the small timescale, which, in turn, is dictated by the period itself. With the method of defect corrected averaging engineered by the authors, it is possible to reach a solution in large time intervals with significantly less calculation time thanks to the use of longer increments. In the following sections, this methodology is introduced when applied to ordinary differential equations of the form

$$\mathbf{y}'(t) = L\mathbf{y}(t) + \mathbf{f}(t), \quad y(0) = y_0 \tag{9.11}$$

In this equation, Ly results from the spatial discretization of an elliptic differential operator, and the discretized source $\mathbf{f}(t)$ is periodical with period ε, that is, $\mathbf{f}(t+\varepsilon) = \mathbf{f}(t)$.

A promising approach consists of replacing the source terms by averaging over a period. Simple averaging, however, leads to solutions that are relatively inaccurate. Defect corrected averaging, in turn, is similar to the concept of stroboscopic averaging by Calvo et al. (2011) and defines a replacement problem, which can be solved with large time increments. The replacement problem is defined so that its exact solution in the stroboscopic points $t = n\varepsilon$ largely conform to the exact solution of the initial problem. The discretized differential operator L remains unchanged here, and the right side $\mathbf{f}(t)$ is replaced by a constant $\mathbf{v}$.

The strategy is illustrated in Fig. 9.4. The highly oscillating solution of the initial problem is marked by circle symbols in the stroboscopic points. The solution by means of simple averaging is a coarse approximation in the stroboscopic points (triangle symbols). The solution of the replacement problem approximates the solution of the initial problem in the stroboscopic points at high accuracy (Fig. 9.5).

Let $S(t_0, t_1, \mathbf{y}_0, \mathbf{f})$ be the solution operator of the differential equation (9.11) with the source $\mathbf{f}(t)$, meaning that the solution in t_1 yields $\mathbf{y}(t1) = S(t0, t1, \mathbf{y}_0, \mathbf{f})$ for

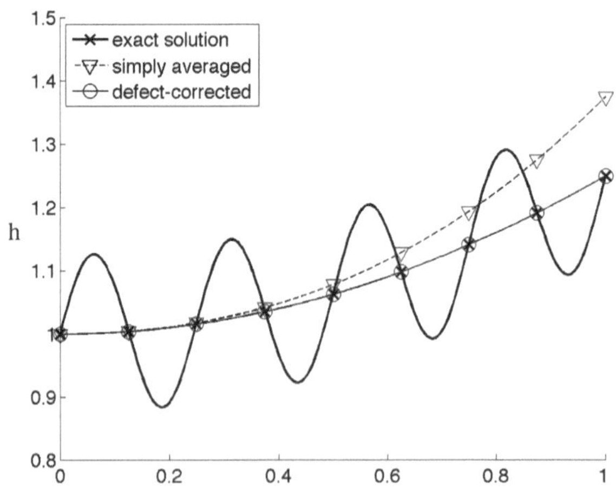

Fig. 9.4 Principle approach in defect corrected averaging: the exact solution (*bold*) is significantly better approximated in the stroboscopic points (*circle symbol*) by defect corrected averaging (*cross symbol*) rather than by simple averaging (*triangle symbol*)

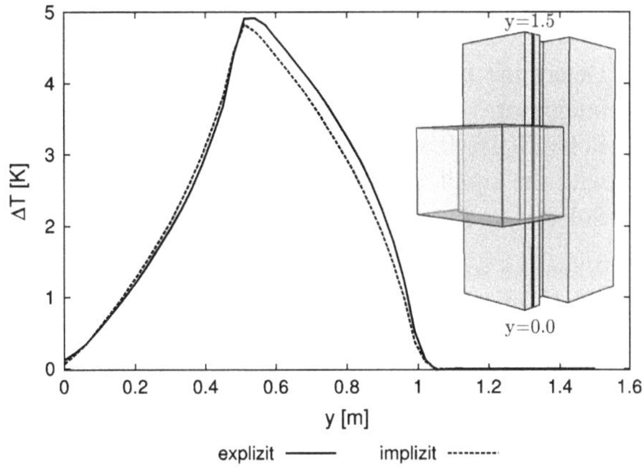

Fig. 9.5 Comparison of reference grid temperature with the diffuse domain solution along the line on Ω_{fix}

$\mathbf{y}(t0) = \mathbf{y}_0$. For a constant source term $\mathbf{f}(t) = \mathbf{v}$, the solution is represented (see Strehmel and Weiner 1982) as

$$S(t_0, t_0 + t, y_0, \mathbf{f}) = \exp(tL)y_0 + t\phi_1(tL)\mathbf{v}, \text{with } \phi(z) = (e^z - 1)/z. \tag{9.12}$$

$$S(0, n\varepsilon, y_0, \mathbf{v}) = S(0, n\varepsilon, (n+1)\varepsilon, y_0, \mathbf{v})\forall y_0, \forall n \in \mathbf{N}. \tag{9.13}$$

is required for the constant replacement source $\mathbf{f}(t) = \mathbf{v}$. Utilising the following properties of the solution operator

$$S(0, t_1, \mathbf{y}_0, \mathbf{f}(t)) = S(0, t_1, \mathbf{y}_0, 0) + S(0, t_1, 0, \mathbf{f}(t)) \tag{9.14}$$

$$S(0, t_1, 0, \mathbf{f}_1 + \mathbf{f}_2) = S(0, t_1, 0, \mathbf{f}_1) + S(0, t_1, 0, \mathbf{f}_2). \tag{9.15}$$

Equation (9.13) is reduced to

$$S(0, \varepsilon, 0, \mathbf{v}) = S(0, \varepsilon, 0, f(t)), \tag{9.16}$$

whereby only $S(0, n\varepsilon, 0, \mathbf{v}) = t\phi_1(tL)\mathbf{v}$ is valid exactly.

A more in-depth consideration of this relation discloses the structure of the ordinary differential equation (ODE). The solution of the ODE with the periodic source over a time interval ε is on the right side. On the left side, we find a solution with the unknown source $\mathbf{v}$. On the left side, a linear operator affects the unknown $\mathbf{v}$. The coefficients of the corresponding matrix are unknown, but the matrix-vector product can be evaluated by solving the ODE numerically over an interval of ε. In other words: A linear equation system $A\mathbf{x} = \mathbf{b}$ has to be solved, in which the coefficients of A are unknown, whereas the products $A\mathbf{x}$ can be calculated. This is the standard situation for the application of (in which .. are used) Krylov solvers. The Krylov method GMRES according to Saad and Schultz (1986) is used here.

The algorithm to calculate a solution in $t = t_e = n\varepsilon$ is described as follows:

1. Determine $\mathbf{b} = S(0, \varepsilon, 0, \mathbf{f})$ with sufficient accuracy by means of a small-scale integrator.
2. Solve $S(0, \varepsilon, 0, \mathbf{v}) = \mathbf{b}$ by means of GMRES, whereby $S(0, \varepsilon, 0, \mathbf{v})$ is calculated with the small-scale integrator.
3. Solve $y' = Ly + \mathbf{v}$ by means of a suitable large-scale integrator in $[0, t_e]$.

When the solution is to be computed not only at the final point $t_e = n\varepsilon$, but also at other stroboscopic points $t_0 \in (0, t_e)$, then it is possible to choose the incremental steps of the large-scale integrator. If the solution also has to be found for non-stroboscopic points t_i, then it is necessary to determine the solution by means of the large-scale integrator at the first previous stroboscopic point, and to subsequently achieve the solution at t_i, by means of a small scale integrator. The method is very efficient, if the solution is sought at a few selected points on a large time scale.

As practical tests show, it is possible to enhance the convergence of the GMRES iteration when initialising with an averaging of the sources over a period so that it is only necessary to correct the defect between the exact solution and the solution which has been achieved through simple averaging over a period. Tests for this simple defect corrected averaging by Naumann and Wensch (2013) enhance defect reduction by a factor of 100 in comparison with simple averaging. Preconditioners, which offer an opportunity for further acceleration of computation, are introduced in Sect. 9.2.4.2.

9.2.3.2 Preconditioning in Defect Corrected Averaging

Our problem consists in the solution of the linear equation system

$$A\mathbf{x} = b \tag{9.17}$$

whereby matrix A results from the spatially discretized differential operator L through

$$A = \varepsilon\phi_1(\varepsilon L) = \varepsilon\frac{\varepsilon L - 1}{\varepsilon L}. \tag{9.18}$$

A good preconditioner P is an approximation of the inverse of matrix A, whereby the condition of P should be significantly less than the condition of A. For this purpose, the authors employ Pade approximations $R(z)$ of the function $\phi_1(z)^{-1}$ in $z = 0$, but, additionally, we demand the same asymptotic behaviour. The resulting preconditioner is $P = R(L)$. Since ϕ^{-1} shows for $z \to -\infty$ the same asymptotic behaviour as $-z$, then $p = q + 1$, with p as the degree of numerator and q as the degree of denominator. This results in $2q$ degrees of freedom, one of which is used for the asymptote at $-\infty$, and the others for the adjustment of the derivates in $z = 0$. This means that one can adjust $2q - 1$ derivates in $z = 0$.

The variants $q = 0$ and $q = 1$ yield:

1. R linear, $R(0) = \phi_1(0)^{-1}$

$$R(z) = -\frac{z}{2} + 1.$$

2. R is a rational fraction, degree of numerator 2, degree of denominator 1

$$R(z) = \frac{z^2 - 2z + 2}{-z + 2}$$

These preconditioners make the defect corrected averaging significantly more efficient and allow for an exact solution for large oscillating ODEs on large time scales.

9.3 Results

9.3.1 Comparison of the Diffuse-Domain Method with Explicit Formulation of the Contact

The two approaches described under Sects. 9.2.2.1 and 9.2.2.2 were compared in terms of computation accuracy and run time characteristics. First, the convergence behaviour of the explicit formulation of the contact was explored. To do this, 3 adaptively refined grids were used. The basic grid in Fig. 9.6 was refined along the heating boundary so that the grid width is halved in this area. To calculate the

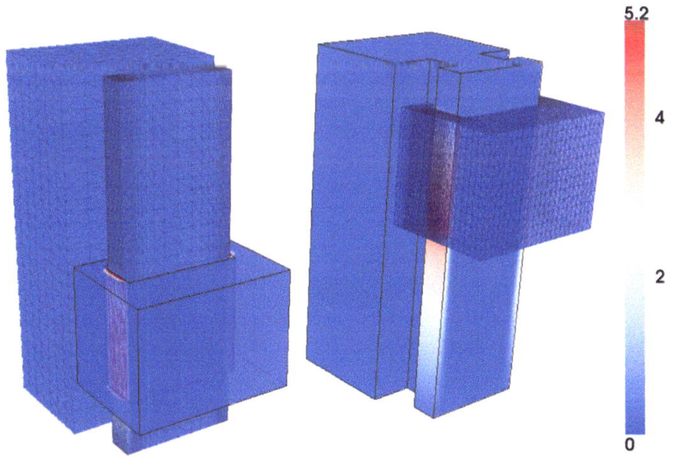

Fig. 9.6 Temperature on the explicit grid with the formulation of the contact at the times 6 and 20 s

Table 9.2 Accuracy and convergence behaviour of the explicit solution at time $t = 12$

$\|\|T_{\delta t,h} - T_{ref}\|\|_2$	h	h/2	$EOC_{\delta t}(h)$
24/80	0.005	0.003	0.6
24/160	0.003	0.002	1.1
24/320	0.003	0.001	1.6
$EOC_h(\delta t)$	0.5	1	1.9

reference solution, the time stepsize $\delta t = 24/320$ and the basic grid refined twice from Fig. 9.6 were used. The temperatures were compared at the point of t = 12 s. At this point, the spindle head is in the half-height position. The accuracy of the explicit simulation is shown in Table 9.2.

The solutions of the explicit discretization are compared not only with the time stepsizes 24/80, 24/160 and 24/320, but also with the basic grid and the grid refined once with the reference solution in the L^2 norm. For sufficiently fine grids, the defect behaviour, shown in Table 9.2, corresponds to the theoretical prediction of the implicit Euler method used. On the basic grid, however, the spatial defect dominates.

The convergence behaviour of the diffuse domain method has already been explored in Franz et al. (2012) and shows quadratic convergence in ε.

For validation, the one-time motion of the spindle head upward is simulated. After each time step, the implicit grid is adapted to the moved geometry. After 81 time increments, the spindle head reaches the upper end of the machine column. Figures 9.7 and 9.6 show the temperature distribution on the subassembly's surface by means of the diffuse domain method and explicit discretization at two different simulation times. The parameters $\rho = 7.200$ kg/m^3, $C_p = 460$ kJ/(kg K),

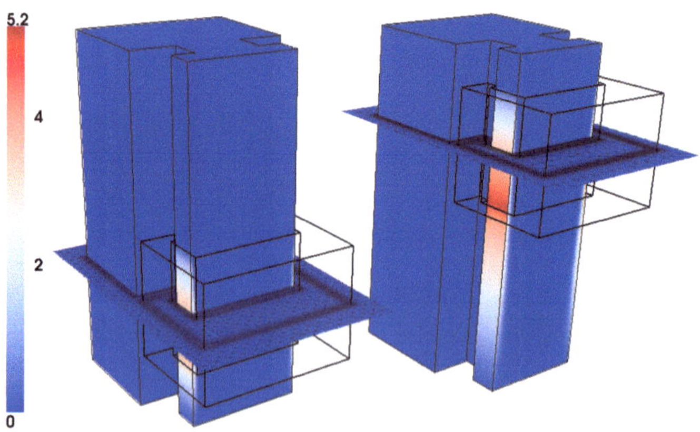

Fig. 9.7 Temperature field on the implicit grid by means of the diffuse domain method (sectional plane) the times 6 and 20 s

$k = 50,000$ kW/(km), $\alpha = 100$ W/K, $\theta = 17.1875$ J and $v = 0.041667$ m/s were used. The interface width was $\varepsilon = 0.2$ m.

The reference solution for the explicit coupling and the diffuse domain solution described were used to compare both methods. Temperature is shown in Fig. 9.5 along the black line on the frictional boundary. Both temperature curves are qualitatively similar. The offset is approximately equal to the washing out due to the phase field width ε.

The defect in the L^2-norm over the entire area is 0.0065 and thus in the order of magnitude of the explicit discretization's numerical error, which confirms the usability of the diffuse domain method. It took about 36 h to carry out the simulation with the diffuse domain method, but only 72 min to perform the explicit formulation of the contact for 320 time increments. The long runtime requirements are caused by the substantially higher number of degrees of freedom. The significant advantage of the approach, which is founded on rapid and simple grid generation and the usability of standard routines, is thus nullified again by the increased runtime demand again. This becomes more obvious when using the original geometry and process conditions. In the following, the explicit formulation of the contact boundary condition was employed.

In both paradigms, the finite element discretization was performed by means of the parallel adaptive finite element library AMDiS by Vey and Voigt (2007), Voigt and Witkowski (2012), however, were used in the sequential mode.

9.3.2 Results of Defect Corrected Averaging

The method was used in order to simulate the thermo-elastic behaviour of the column/spindle configuration in its complex geometry over a period of 15 h. This problem was regarded as the reference problem to evaluate and benchmark the MOR techniques developed under A05 and A06. The periodic source is given here by the frictional heat at the periodically moved contact position. For the whole boundary, heterogeneous Robin boundary conditions are used, which represent the ambient cooling. In this time span, 75–2,400 periodical motions are executed per hour. The solution calculated by means of the small scale indicator was used as reference solution. The comparison involves the solutions with simple averaging and defect corrected averaging, where the latter was used with preconditioner. The results in Table 9.3 demonstrate that defect correction enhances simple averaging.

Table 9.3 Runtime versus accuracy—compared for different averaging methods

Method	Runtime	Error (K)
Small scale integration	72 h	
Simple averaging	3 min	0.25
Defect corr. averaging with preconditioning	22 min	0.025

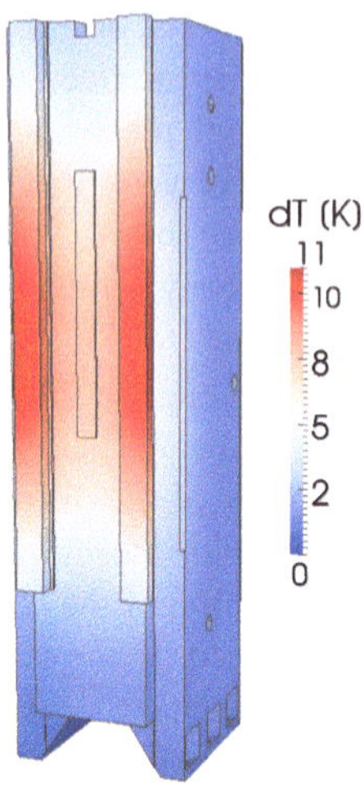

Fig. 9.8 Temperaturfeld mit defektkorrigierter Mittelung nach 15 h

The cost of this advantage is an increased calculation time; however, this is still lower by some orders of magnitude than in small scale simulation. In this simulation, one third of the correction effort results from the application of the preconditioned GMRES, and the other two thirds from the calculation of the right side with 100 steps. After simulation of the thermal behaviour, the temperature fields are employed to determine the corresponding deformations from linear elasticity.

The combination of the two approaches permits an efficient simulation of the machine tool with its moving subassemblies. Figures 9.8 and 9.9 illustrate the calculated temperature fields and the amount of deformation in the original geometry with the given, varying periods after 15 h. The calculation is also carried out in AMDiS.

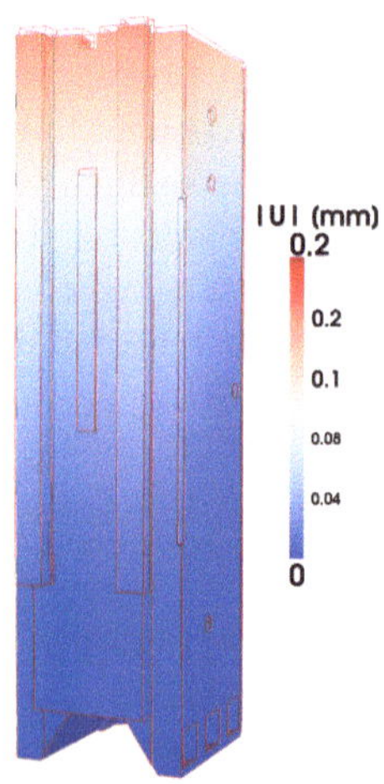

Fig. 9.9 Absolutwert der Verschiebungen nach 15 h. Die roten Linien entsprechen der deformierten Geometrie mit 100 facher Verstärkung der Verschiebungen

Figure 9.9 clearly shows the strongest heating at the frictional boundary; the difference in temperature is here 11 K. In the correspondent right Fig. 9.9, the absolute displacement is shown. The red lines represent the deformed geometry, with the deformation amplified by a factor of 200 for visualisation. A detailed comparison of the simulation results with the subprojects A05 and A06 and the measurements gained in experiments is now underway.

9.4 Classification in the CRC/TR 96

In subproject A07, different mathematical methods for the simulation of the thermo-elastic machine tool behaviour were investigated. Coupling strategies for finite element spaces moved against each other were developed to simulate thermo-elasticity in moving geometries. The interpolation algorithm applied here provided exact results and is used to generate reference solutions for other subprojects. As an alternative, the diffuse domain approach was tested. This approach provides similar results and permits the use of standard methods and an easy processing of geometries, but suffers, however, from substantially longer computer runtime. The high resolution reference solutions generated contribute to the validation of the numerical results obtained using the commercial software packages in A05 and the results of the model reduction methods in A06. The comparisons (benchmarks) will be published elsewhere.

9.5 Outlook

Defect corrected averaging was developed and further refined by preconditioning techniques for the simulation over long time intervals, in which periodic motions appear over large partial intervals. This method makes possible efficient simulation over very long time intervals and is significantly more accurate than simple averaging techniques. An additional strength is that it is substantially faster than a simulation on the small time scale of periodic motions. After successful processing of spatial discretization, the focus is on further improvement of time discretization. At the time of publication, the method is being expanded by the Robin boundary conditions that they are needed in subproject A05. In the next phase, temporal integration methods that can be paralleled will be used in order to carry out simulations more efficiently over greater time spans. The methods of choice in terms of the structure of the thermo-elastic FE models are the approaches by ParaReal (Nielsen 2012), as well as ParaExp (Gander and Güttel 2013).

References

Calvo MP, Chartier P, Murua A, Sanz-Serna JM (2011) Numerical stroboscopic averaging for ODEs and DAEs. Appl Numer Math 61(10):1077–1095. doi:10.1016/j.apnum.2011.06.007. http://dx.doi.org/10.1016/j.apnum.2011.06.007

Franz S, Gärtner R, Roos H, Voigt A (2012) A note on the convergence analysis of a diffuse-domain approach. Comp Meth Appl Math 12:153–167

Gander MJ, Güttel S (2013) Paraexp: A parallel integrator for linear initial-value problems. SIAM J. Scientific Computing

Li X, Lowengrub J, Raetz A, Voigt A (2009) Solving pde's in complex geometries: a diffuse domain approach. Commun Math Sci 7:81–107

Naumann A, Wensch J (2013) Defect corrected averaging for highly oscillatory pdes. PAMM (accepted for publication)

Nielsen AS (2012) Feasibility study of the parareal algorithm. Master's thesis, Technical University of Denmark, DTU Informatics, E-mail: reception@imm.dtu.dk, Asmussens Alle, Building 305, DK-2800 Kgs. Lyngby, Denmark. http://www.imm.dtu.dk/English.aspx, DTU supervisor: Allan P. Engsig Karup, apek@imm.dtu.dk, DTU Informatics

Saad Y, Schultz M (1986) GMRES: a generalized minimal residual algorithm for solving nonsymmetric linear systems. SIAM J Sci Stat Comput 7:856–869

Strehmel K, Weiner R (1982) Behandlung steifer Anfangswertprobleme gewöhnlicher Differentialgleichungen mit adaptiven Runge-Kutta-Methoden. Computing 29:153–165

Vey S, Voigt A (2007) AMDiS: adaptive multidimensional simulations. Comput Vis Sci 10 (1):57–67. http://dx.doi.org/10.1007/s00791-006-0048-3

Voigt A, Witkowski T (2012) A multi-mesh finite element method for lagrange elements of arbitrary degree. J Comput Sci 3:420–428

Chapter 10
Modelling of Thermal Interactions Between Environment and Machine Tool

Reimund Neugebauer, Steffen Ihlenfeldt, Carsten Zwingenberger, Janine Glänzel and Carsten Richter

Abstract Thermal boundary conditions, such as heat flux, play an important role in the overall context of representing machine tools' thermal behaviour and their interactions with the environment in a way that is close to reality. In this contribution, the machine tool–environment interface is represented based on the finite-element method and the required parameters are assigned. The term "environment" refers in this paper to the surrounding atmosphere and sometimes the foundation. The approach includes the determination of the relevant environmental boundary conditions, such as convection and heat radiation, and the description of how to integrate them systematically into the finite-element model.

10.1 Introduction

Comparing measurements and simulations, it can be seen that the computational analysis of the machine tools' thermal behaviour and performance evinces an obvious potential for improvement potential concerning their reliability in terms of quality and quantity (Neugebauer et al. 2009; Großmann et al. 2013). Particularly in the field of the coupled analysis of thermal properties and properties from structural mechanics, there is a lack of sufficiently exact knowledge of the description of interactions between the environment and the machine tool. As a rule, this limitation arises from difficulty in describing the environmental conditions as precisely

R. Neugebauer (✉) · S. Ihlenfeldt · C. Zwingenberger · J. Glänzel · C. Richter
Fraunhofer Institute for Machine Tools and Forming Technology IWU, Chemnitz, Germany
e-mail: reimund.neugebauer@iwu.fraunhofer.de

S. Ihlenfeldt
e-mail: steffen.ihlenfeldt@iwu.fraunhofer.de

J. Glänzel
e-mail: janine.glaenzel@iwu.fraunhofer.de

C. Richter
e-mail: carsten.richter@iwu.fraunhofer.de

K. Großmann (ed.), *Thermo-energetic Design of Machine Tools*,
Lecture Notes in Production Engineering, DOI 10.1007/978-3-319-12625-8_10

as necessary. The values found in simulation—for heat flux density, heat transfer coefficients and heat emission parameters—frequently differ by more than 50 % from those measured. This phenomenon is, for instance, based on the fact that in transient thermo-elastic simulations, the use of homogeneous environmental temperatures results in convection and radiation conditions that are insufficiently exact. Finally, the complex feedback between the thermal effects of conduction, convection, and heat radiation, as well as their non-linear characteristics, complicates a reliable mapping of the different influencing factors. However, the most exact possible prediction of the machine properties directly depends on the quality of the input data that provide the boundary conditions for the simulation models.

Concerning the representation of interaction between structure and environment due to convection, empirical calculation guidelines exist for simple geometric bodies given defined assumptions (VDI 2006), which have to be chosen for each real case of application on a machine tool. However, this analytical strategy of determination of heat fluxes does not provide a sufficiently exact prediction of complex thermal machine tool properties. Consequently, in this subproject, boundary conditions found by calculations are also used. In addition to the analytical calculation, fluid simulation is employed as a new approach to determine the thermally relevant parameters for convection, heat transfer and heat radiation.

10.2 Approach

In an innovative strategy, minimal models are generated for the modelling of boundary conditions, which become a part of the structural mechanical finite element (FE) simulation to calculate the thermo-elastic displacements of an assembly. To determine the relevant boundary conditions for convection, not only analytical computations are carried out, but also CFD simulations by means of ANSYS-CFX. For this purpose, it is necessary to model the corresponding FE meshes to map the flow conditions around the assembly. The flow model applied first considers a laminar flow for immobile assemblies, such as the machine column in standstill position. Based on this, the moving assemblies (such as the moving machine column) are investigated, and this makes it necessary to represent forced flows.

A FE model needed to verify the machine tool to be explored has to be generated at the beginning. The minimal models of complex assemblies, such as mounting systems and ball screws, as they are described here and to be developed in the other subprojects, will be integrated into this model later on. The parameters for the minimal models are identified to represent the heat transfer behaviour. The parameters found are introduced afterwards in the mechanical FE model for the thermo-elastic displacements' calculation. This methodology is verified by means of examples from industrial practice. The options for an automated definition of the thermal environmental boundary conditions in the application of the FE method, which provides a uniform calculation algorithm and the integration of the known empiric calculation equations in order to express the interaction between the machine tool structure and its environment, are enhanced. The procedures describing the transfer of heat are mapped by the three transfer mechanisms of thermal conduction, convection, and radiation (Baehr and Stephan 2004). The temperature field thereby derived shapes the basis for the calculation of the thermally induced displacements. The consequences of the resulting displacements of the tool centre point (TCP) directly affect the machine tool's machining accuracy and are thus significant when deducing and evaluating design measures or correction methods.

10.3 Results

10.3.1 Modelling of Thermal Influences and Interactions

The thermal convection and radiation transfer mechanisms mostly occur in parallel; however, they are sometimes linked to different conditions. The amount of the transferred heat flux due to radiation strongly depends on the temperature level. This is caused by the fact that when calculating the heat fluxes due to thermal radiation, absolute temperature enters in the fourth power, whereas we have a linear relationship in convection. Because the surfaces that are involved in the transfer by radiation are sometimes very large, in every case it must be determined whether thermal radiation should also be taken into account in the representation of the boundary conditions even at room temperature.

Heat transfer procedures between the structure and the environment were evaluated in terms of their relevance based on fundamental investigations and comparisons of thermal radiation and convection, which are, in turn, based on analytical calculation methods and the use of numerical approaches, as well as sensitivity analysis executed for an assembly chosen as an example.

10.3.1.1 Representation of the Boundary Conditions for Convection

In addition to analytical calculation methods, various numerical methods are available to calculate the energy transfer by thermal convection.

Numerical Versus Analytical Calculation

Analytical method For solids with simple geometries, it is possible to analytically calculate the heat fluxes transferred to the environment by means of convection. Here, particularly in the case of free convection, simplifying assumptions have to be employed to make the known empirical computation equations usable. Thus, for instance, a plane plate A_1 of 1 m^2 surface is chosen. To calculate the heat transfer coefficients with free convection, empirical equations from (VDI 2006).

$$Q_{Konv} = h \cdot A_1 \cdot (T_1 - T_2) \tag{10.1}$$

The heat flux transferred between the machine tool structure and environment is, according to Eq. (10.1), linearly dependent on the temperature difference $\Delta T = T_1 - T_2$ and the heat transfer coefficient h to be defined following Groth and Müller (2001).

Numerical method Temperature fields inside a structure can not only be calculated based on structural mechanical (mechanics) FE models, but also by means of Computational Fluid Dynamics (CFD). This does not only provide the option to analyse the temperature field inside the structure and the fluid, which is also involved in the model; the heat flowing on the boundary between the structure and fluid also permits an evaluation of the heat transfer coefficient. This qualifies this kind of numerical analysis to be applied to commensurable considerations of the boundary conditions found empirically and thus to detect existing deficits in the description of convection.

When comparing the two calculation approaches by tendency, it was possible to ascertain that the analytical determination of the heat flux between the surfaces and the environment predicts a more intensive heat transfer than forecasted by CFD simulation. Despite the relatively simple geometries analysed, the two procedures differ by maximally 35 % and thus indicate a clear need for action. It can be assumed that the "high-order" CFD simulation model provides more accurate results than the analytical procedure based on empirical equations. However, the considerations introduced do not permit us to conclude beyond doubt whether the trend towards the prediction of a higher heat transfer in the case of analytical calculation is also true for real component geometries. Figure 10.1 illustrates the higher growth of the heat fluxes in the direct comparison of the analytically and numerically computed heat fluxes, shown for a horizontal plate.

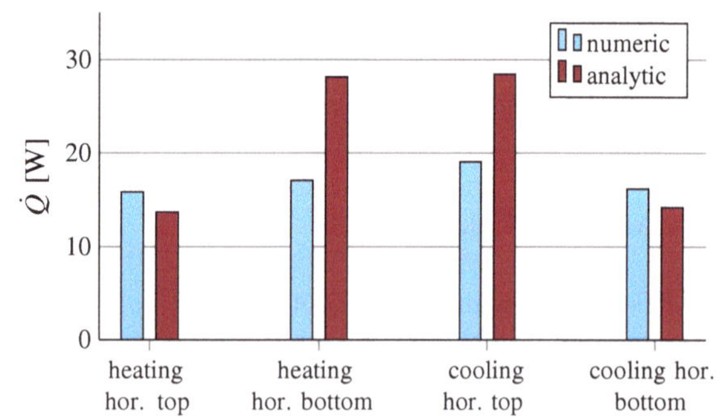

Fig. 10.1 Heat fluxes due to convection on a horizontal plate obtained from numerical and analytical calculations (high mesh density in the boundary layer)

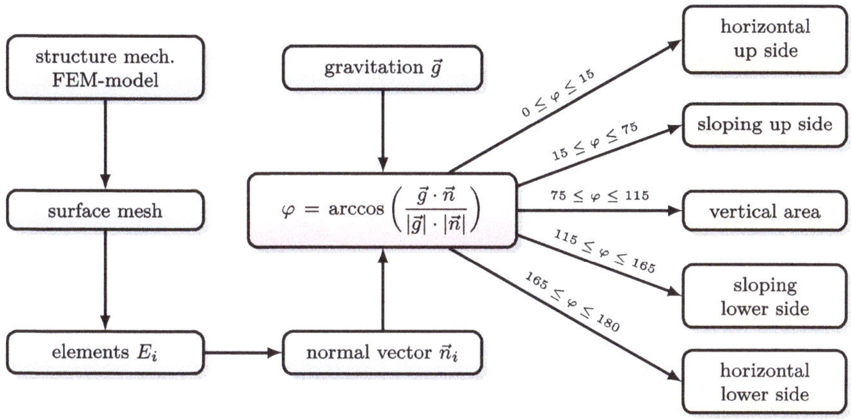

Fig. 10.2 Classification of the structural surface—algorithm

Systemised definition of heat transfer coefficients The heat flux transferred into the environment depends on two influencing variables—the difference in temperature between the structure and the ambient air—and the heat transfer coefficient, which, in turn, depends on several factors. One of them is also the difference in temperature so that the temperature field, which also changes, results in heat transfer coefficients of the problem that change as a function of time. Starting from the structural mechanical simulation model, the surface elements of the structure are identified according to their geometric properties (such as horizontal surface 0–15°, vertical surface of 75–105°, …), which permits a classification into five different regions. In order to make this approach viable for real component geometries, it is necessary to define limits for the angle of inclination for the five areas mentioned above. As a rule, a value of ±15° is introduced for the deviation from horizontal and vertical alignment to distinguish the individual surface orientations. This classification is also used here as a distinctive criterion to express the position to the gravitation vector. The algorithm to classify the surface elements of the structure investigated is elucidated in Fig. 10.2 in a schematic diagram.

The result of this classification obtained for the machine column is shown in Fig. 10.3. The orientation of the elements against gravitation has to be defined, and the elements have to be grouped according to their distance from the foundation. Comprehensive measurements showed a clear influence of the foundation temperature on the air temperatures, particularly on the machine tool mounting, which, in turn, results in air layers of different temperatures.

To determine the individual heat transfer coefficients, analytical calculation rules are applied to the indicated areas, which take into account the geometric prerequisites in terms of free convection, see also (Neugebauer et al. 2010). The temperature stratifications and differences between the structure and the environment are taken into account. Thus, it is possible to determine the heat transfer coefficient in a systemised manner and to transfer (port) the results to the corresponding surface elements.

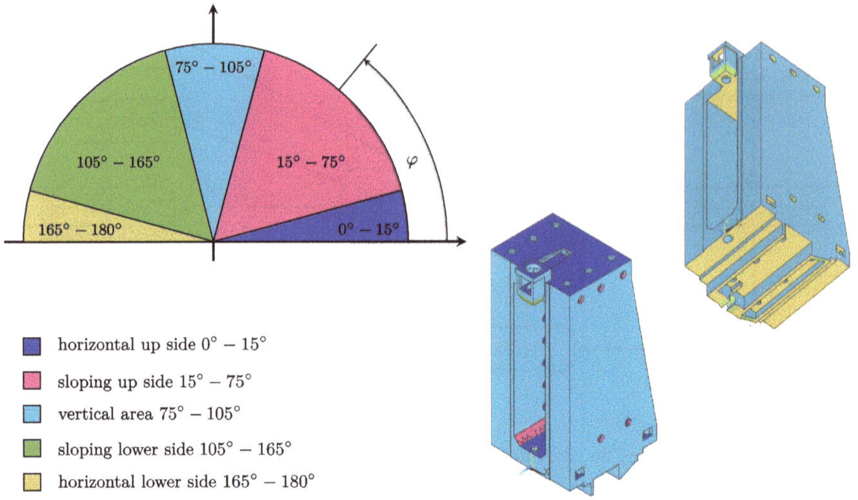

Fig. 10.3 Classification of the structural surface, shown for the machine tool column

As can be seen from the flow simulations for the load case of a machine column heated by an external heat source, different flow velocities are established in the surrounding air. This phenomenon mainly results from the structural formation of the geometry. Figure 10.4 clarifies the so-called "chimney effect" inside the machine tool structure. The sectional view through the surrounding air makes it possible, among other things, to analyse the flow velocities. Factors like these that influence the heat transfer between structure and environment either cannot or can only be incompletely captured by means of empirical computation equations. Flow simulation can provide essential results here, which could otherwise only be identified by experiments with great effort. The transferred heat fluxes are not only affected by the flow velocities mentioned above, but also by the differences in the structural and environmental temperatures, and, consequently, have to be considered. CFD simulations can guarantee that the structural areas affected thereby are correctly identified.

10.3.1.2 Representation of the Radiation Boundary Conditions

Analytical approach The analytical computation of the heat flux from a plane plate to the surrounding air due to radiation functions without the determination of the radiation coefficient. This coefficient is 1.0, if radiation is transferred between the surface and the overall hemi-cube. Thus, another factor is that only the emissivity is entered into the calculation. The transferred heat flux is only influenced by the difference in temperature, the emissivity and the area of the surface. Heat flux is calculated by means of the equations below (VDI 2006), capt. 5, Formula (10.5)

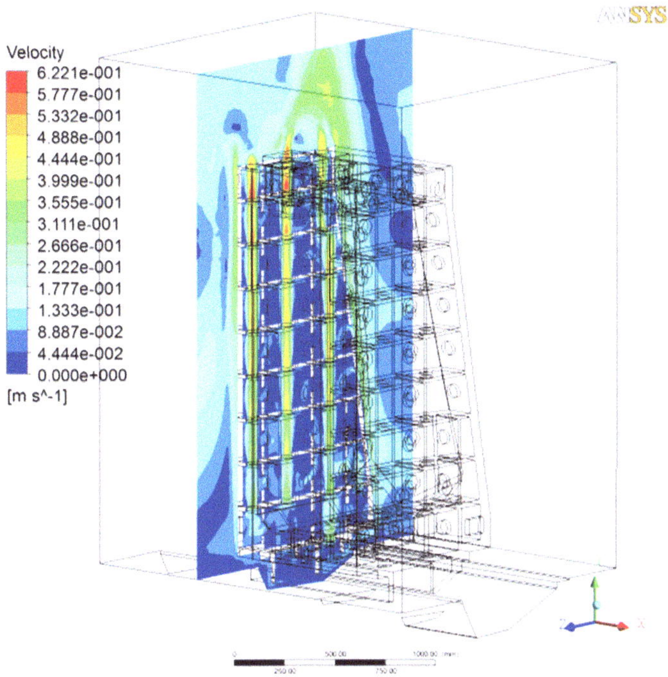

Fig. 10.4 Flow velocity of air around the machine tool atmosphere (free convection)

$$Q_{str} = \sigma_{12} \cdot A_1 \cdot \left((T_1)^4 - (T_2)^4 \right) \tag{10.2}$$

$$\sigma_{12} = \sigma_{s\epsilon_1\epsilon_2} \frac{\nu_{12}}{1 - \frac{(1-\epsilon_1)(1-\epsilon_2)A_1}{A_2} (\nu_{12})^2}$$

Since the heat flux transferred due to thermal radiation into the environment is mostly directed in parallel to the heat flux by convection, it is possible to formulate a heat transfer coefficient h_{str} valid for thermal radiation by means of linearisation of the temperature difference ${T_1}^4 - {T_2}^4$ between the wall temperature T_1 and the fluid temperature T_2. Heat flux according to radiation according to Eq. (10.2)

$$\begin{aligned} Q_{str} &= \sigma_{12} \cdot A_1 \cdot \left((T_1)^4 - (T_2)^4 \right) \\ &= h_{str} \cdot A_1 \cdot (T_1 - T_2) \end{aligned} \tag{10.3}$$

can be calculated by using the binomial formula

$$(T_1)^4 - (T_2)^4 = (T_1 - T_2) \cdot (T_1 + T_2) \cdot \left((T_1)^2 - (T_2)^2\right) \tag{10.4}$$

in a linearised manner.

$$h_{str} = \sigma_{12} \cdot (T_1 + T_2) \cdot \left((T_1)^2 - (T_2)^2\right) \tag{10.5}$$

Numerical approach The FEM software ANSYS applied involves three distinct strategies to calculate radiation effects on three-dimensional models for a numerical analysis based on structural mechanical models:

- Application of surface effect elements to model surface effects
- Use of a radiation matrix in the form of a superelement in a thermal analysis
- Use of a radiosity solver

The method mentioned first, based on surface effect elements, is suitable to model radiation effects between a surface and a point, and was employed for numerical implementation. The methods mentioned second and third, using the radiation matrix and the radiosity solver, can be used in a more general sense and also makes it possible to model the interaction between several surfaces and the environment. It takes more calculation time to create the radiation matrix by means of the radiosity method for large models. A limitation when using the radiation matrix arises from the option of defining temperature-dependent emissivities dropped here; this, however, is not relevant for an application to a machine tool structure.

Numerical versus analytical computation The numerical solution quantifying the radiation to the environment does not appreciably differ from the analytical one. This is made clear by the difference of approx. 16 % obtained for two surfaces that are in a position vertical to one another. Figure 10.5 shows the results gained for both the analytical and numerical solutions of the problem in comparison.

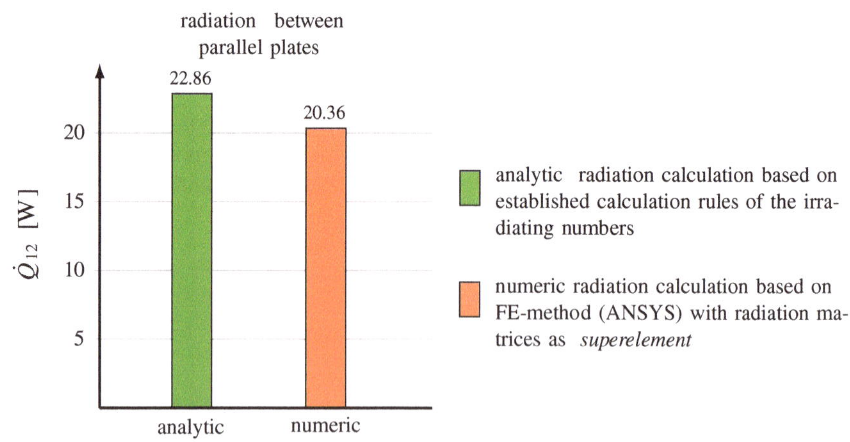

Fig. 10.5 Analytical and numerical computation results in comparison

The introduction of an overall heat transfer coefficient (compare Zwingenberger 2014) $h_{ges} = h_{str} + h_{konv}$ makes it possible to represent the interaction with the environment for convection and thermal radiation in a combined manner and at the same time reduces the calculation time when using the Finite-Element method, since it is not necessary to run additional iterations because of additional nonlinearities.

10.3.2 Sensitivity Analysis on a Machine Tool Structure

The influence of each individual thermal boundary condition in the total heat flux was evidenced by a sensitivity analysis, shown for a column assembly and spindle unit, as well as the required components, consisting of linear guidances, ball screw, roller bearing and feed drive. The model used is based on a prototype-like machining centre by the firm Auerbach Maschinenfabrik GmbH. In Fig. 10.6, the analysed part of the machine structure is shaded. The main spindle of the prototype is represented by a dummy with no machining function; it represents the main spindle in terms of geometry and mass.

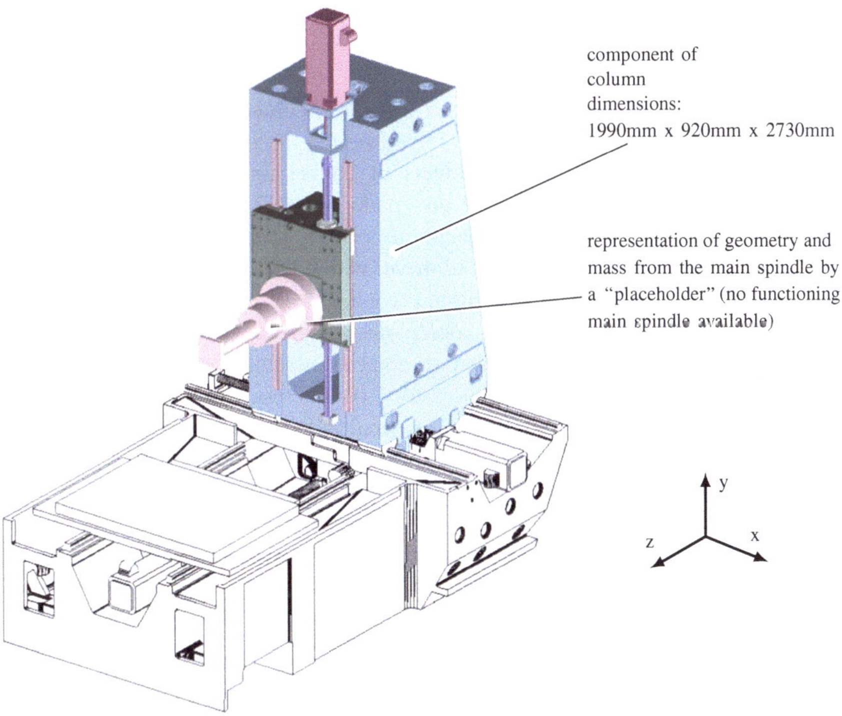

Fig. 10.6 Components of the simulation model for sensitivity analysis (Auerbach ACW630)

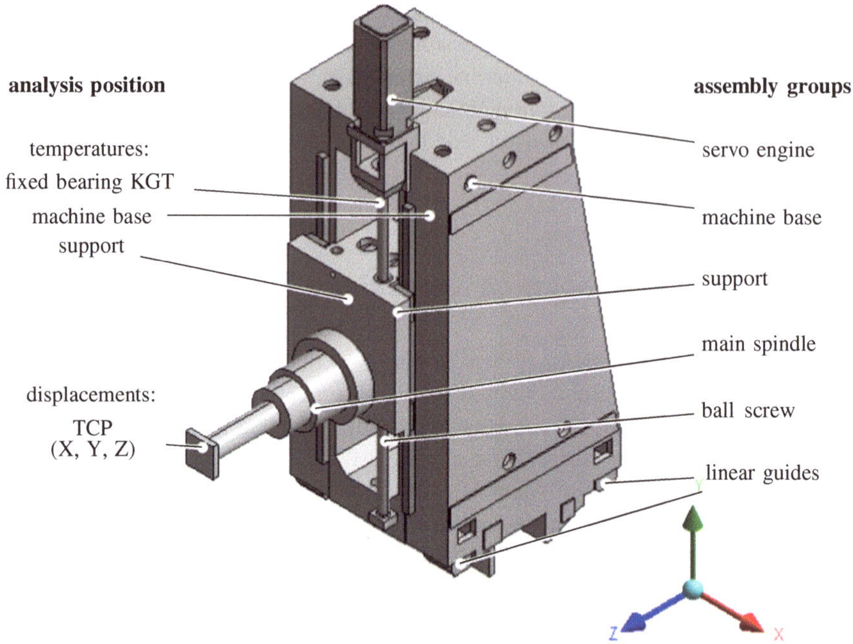

Fig. 10.7 Analysis positions on machine tool column

The partial model of the machine tool structure is illustrated in Fig. 10.7. Since the structure's sensitivity was checked regarding the thermal boundary conditions, that is, a comparison between the effects of convection and thermal radiation, fundamental conclusions in this case are made possible by considering an individual complex assembly from the machining centre. The load case assumed takes into account the heat flux from the feed drive, which brings into action the vertical slide motion. The heat source is assumed to be 250 W independently of the real load cycle; it is defined as a heat source rate in the feed drive. The heat flux dissipating into the environment as well as adjacent assemblies as power loss of the feed drive was derived from the servo-motors' parameters. The splitting of the heat flux from power losses into convection, thermal radiation and thermal conduction on the motor flange largely depends on the mounting conditions, above all on the thermal conduction into the machine tool structure, and cannot be expressed in a generalised way. However, as has been shown by investigations of the thermal losses and their introduction into the machine tool structure, the relevant share is transferred by thermal conduction via motor flange to the adjacent components (Neugebauer et al. 2009). A temperature boundary condition assumed to be 20 °C on the linear guides in X direction defines the interface with the machine bed.

Influences of convection and radiation Figure 10.8 indicates the temperatures under different convection boundary conditions for chosen analysis positions

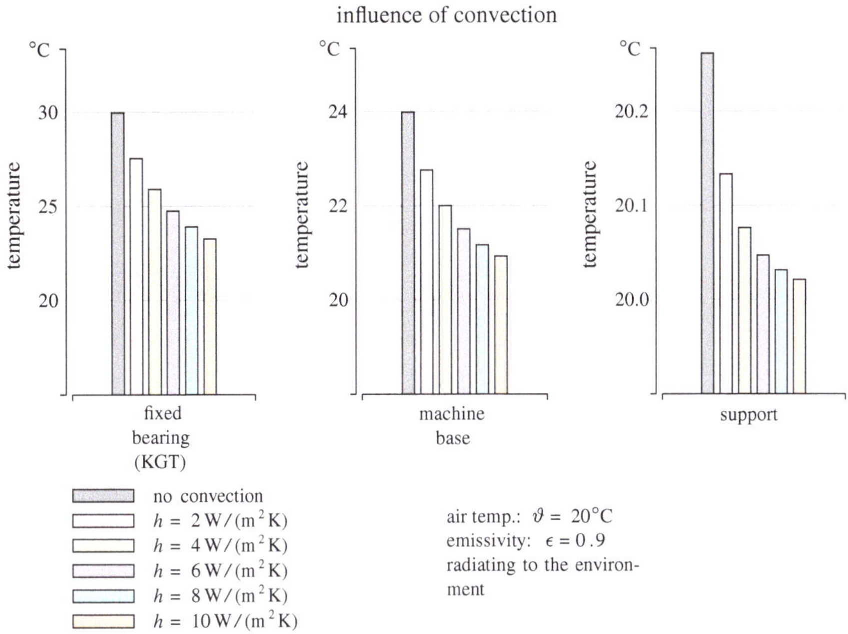

Fig. 10.8 Influence affected by convection on the temperature field (chosen measurement points)

(see Fig. 10.7) on the assembly investigated. The thermal interactions between the structure and the environment are characterised by defined boundary conditions for the thermal radiation (emissivity $\varepsilon = 0.9$) and a constant ambient temperature of 20 °C. There are no interactions between the individual parts of the structural surface, which would permit a heat flux at an assumed temperature difference due to visual contact. The heat transfer coefficient varies from simulation to simulation, but remains constant during the calculation of the stationary state.

On the positions evaluated on the structural surface, the mean variation of the heat transfer coefficient results in a deviation of 22.4 % on the rigid bearing of the ball screw of the vertical axis, and 1.2 % on the slide. Since the temperature difference diminishes as a function of the heat source distance, the heat fluxes between the structure and the environment also decrease, which finally results in a reduced influence of the boundary conditions defined.

The varied boundary conditions for radiation in the sensitivity analysis include the interaction between the structure and the environment. For a comprehensive analysis and a representation of all effective heat fluxes induced by radiation, interaction between the individual assemblies has to be taken into consideration, if they are in mutual visual contact. In the concrete case, for a consideration like this, it is necessary not only to include the structural assemblies, but also the entire machine tool cover. Since the machine tool structure investigated had no cover, and the interaction due to thermal radiation between the assemblies and for an individual

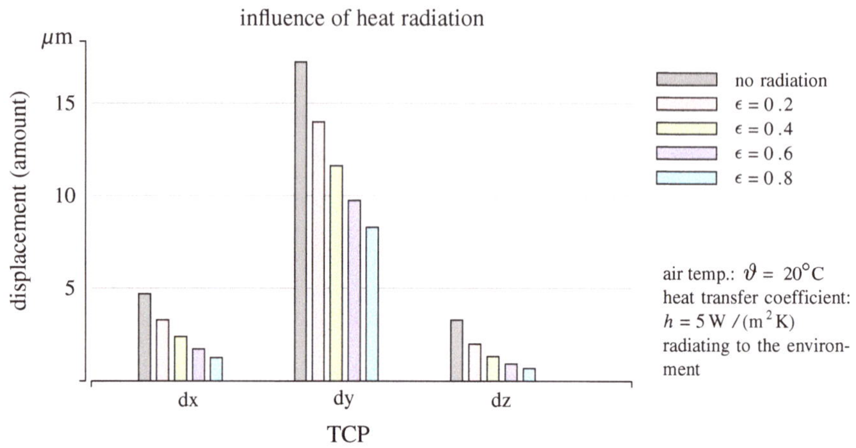

Fig. 10.9 Influence of thermal radiation on the thermally induced TCP displacements

component (in the case of radiation to itself) would cause considerable calculation efforts, sensitivity analysis was limited to radiation into the environment.

The influence of the used radiation boundary conditions used on the simulation outcomes can be taken from Fig. 10.9. The convection boundary conditions acting in parallel are identical for all variants investigated, and defined by a uniform heat transfer coefficient h = 5 W/(m^2 K) and an ambient temperature η = 20 °C, which is also chosen as a constant. Emissivity is defined as a variable parameter between $\varepsilon = 0$ (without thermal radiation) and $\varepsilon = 1$.

Analysis The sensitivity analysis executed on the assembly introduced indicates that both thermal effects explored—free convection and thermal radiation—make up similar percentages of the overall heat flux. Thus, both heat transfer mechanisms have to be considered here with sufficient accuracy, when analysing the interaction between the machine tool structure and the environment, in order to be able to capture the heat flux transferred on the boundary surfaces to the environment in terms of quality and quantity in a manner that is closely to reality.

When expending the load cases considered by the traverse motions of the individual assemblies and the resulting air motions, the share of the heat flux transferred due to thermal radiation is reduced significantly for the surfaces situated in the area influenced by forced air flows. For the part of a machine tool structure without a machine cover, like that studied here, the interaction between parts of the structural surface that are in mutual visual contact is relatively minor. If a machine tool cover exists, then areas of the machine surface are separated from each other, which means that visual contact is avoided and, at the same time, that a percentage of the emitted radiation is reflected by the cover and absorbed by the structural surface. This effect normally also appears inside the machine tool structure, at the interaction between the walls of cavities; however, it remains unconsidered in the chosen definition of the radiation boundary condition to the environment.

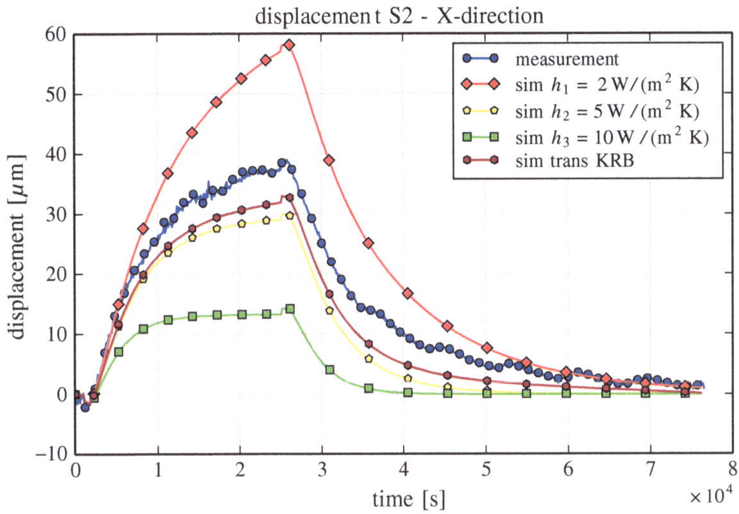

Fig. 10.10 Thermally induced TCP displacement in X axis direction

10.3.3 Verification in Experiment

Figure 10.10 elucidates the displacement on the spindle collar for the X axis direction. The stationary simulation initial state is used as a homing position, which means that the value at the point in time t = 0 is subtracted from all displacement values. Consequently, only the displacements caused by the temporally changed temperature field are outlined and shown.

The TCP displacements are reproduced in a manner close to reality in terms of quality and quantity by means of the simulation with transiently adapted heat transfer coefficients. In general, looking at the thermal displacements calculated, it can be ascertained that the method introduced for transient adaptation of the heat transfer coefficients provides the most reliable prediction.

10.4 Classification According to the Goals of the CRC/TR 96 and Outlook

The activities executed in subproject B01 "Modelling of thermal interaction between environment and machine tool" describe the interface between the environment and the machine tool. The studies and work performed to determine the influences of convection and radiation were able to establish the scientific basis for improved modelling of the thermal interactions between the environment and the machine tool, shown for a machine column. The authors succeeded in classifying surface areas by means of structural analysis and the assignment of individual heat transfer coefficients. Based on these outcomes, calculation models able to map the complete machine tool with all external surfaces and its geometrical features can be established. For this purpose, approximation functions allowing structural mechanical simulations to be used without CFD simulations have to be developed. These approximation functions contribute to an optimal transfer of the parameters obtained from flow simulation to the surface of the FE model. Based on these results, the structural mechanical FE model can be developed and calculated efficiently, reducing the calculation time. This makes it possible to calculate variants with different alpha values and to optimise parameters.

Initial preliminary studies on sensitivity related to the machine tool surface geometry and the relevant boundary conditions were executed to develop this approximation functions and to implement it as a software module. Based on these outcomes, the high quality and efficiency of the approximation functions to be developed have to be assured in connection with performance data analyses. In the first stage of application, the analyses are mainly carried out for free flows.

References

Baehr HD, Stephan K (2004) Wärme- und Stoffübertragung, Auflage: 4., neu bearb. u. erw. Aufl. Korr. Nachdruck. Springer, Berlin u.a

Großmann K, Müller J, Merx M, Riedel M (2013) Grundlagen der experimentellen Analyse des thermo-elastischen Verhaltens von Werkzeugmaschinen mit Hilfe der selektiven Thermografie. ZWF Zeitschrift für wirtschaftlichen Fabrikbetrieb 118:492–497

Groth C, Müller G (2001) FEM für Praktiker - Band 3: Temperaturfelder: Temperaturfelder: Basiswissen und Arbeitsbeispiele zu FEM-Anwendungen der Temperaturfeldberechnung - Lösungen mit dem FE-Programm ANSYS(R). expert, Renningen-Malmsheim

Neugebauer R, Ihlenfeldt S, Zwingenberger C (2010) An extended procedure for convective boundary conditions on transient thermal simulations of machine tools. Prod Eng Res Dev 4:641–646. doi:10.1007/s11740-010-0263-0

Neugebauer R, Zwingenberger C, Schädlich K, Vesely J (2009) Simulation des Wärmegangs von Werkzeugmaschinen, Chemnitz

VDI-Gesellschaft Verfahrenstechnik und Chemieingenieurwesen (2006) VDI-Wärmeatlas Berechnungsblätter für den Wärmeübergang, Karlsruhe

Zwingenberger C (2014) Beitrag zur Verbesserung der Simulationsgenauigkeit bei der Bestimmung des thermischen Verhaltens von Werkzeugmaschinen. Dissertation, TU Chemnitz

Chapter 11
Determination and Modelling of Heat Transfer Mechanisms Acting Among Machine Tool Components

Sarah Vieler, Yona Frekers, Michael Burghold and Reinhold Kneer

Abstract Quantifying the thermal resistances between machine components in contact with each other has key importance for the characterisation of the machine tool's thermal behaviour. Against this background, this paper outlines analytical, experimental and numerical approaches to calculate thermal resistances at contacting surfaces.

11.1 Introduction

The aim of the CRC/TR 96 is to create an overall model for transient thermo-elastic machine tool behaviour. For this, many individual machine tool components are modelled by means of finite-element-method (FEM). In order to achieve a reliable prediction of the thermo-energetic behaviour, heat flows throughout various components, as well as at interfaces between contacting components need to be taken into account. These interfaces include rigid joints at the spindle, tool chuck or machine table. However, heat is also transferred between components under relative movement, such as in guiding elements. Figure 11.1 provides a schematic overview: Three components are in contact with each other, whereas the left is in a fixed coupling with the middle component. The right component is moved relatively to the middle one, resulting in friction and thereby imposing a heat source. For this reason, heat flows throughout the assembly occur.

S. Vieler · Y. Frekers (✉) · M. Burghold · R. Kneer
Institute of Heat and Mass Transfer, RWTH Aachen University, Aachen, Germany
e-mail: frekers@wsa.rwth-aachen.de

S. Vieler
e-mail: vieler@wsa.rwth-aachen.de

M. Burghold
e-mail: burghold@wsa.rwth-aachen.de

R. Kneer
e-mail: kneer@wsa.rwth-aachen.de

K. Großmann (ed.), *Thermo-energetic Design of Machine Tools*,
Lecture Notes in Production Engineering, DOI 10.1007/978-3-319-12625-8_11

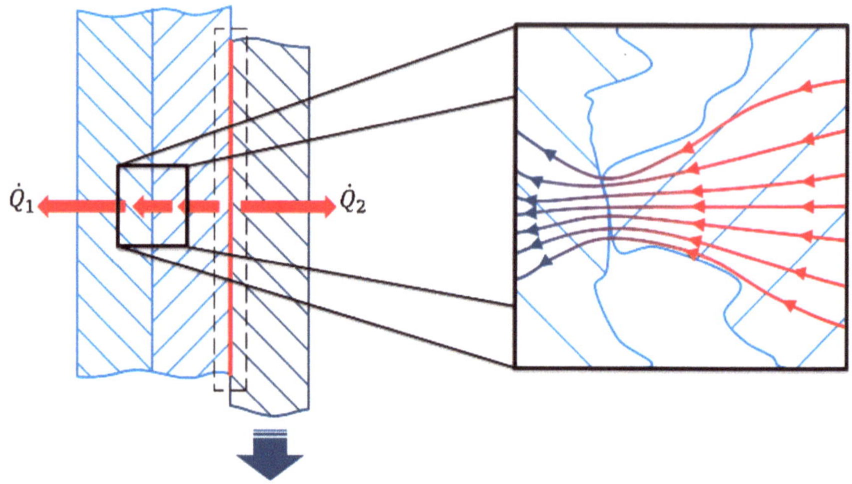

Fig. 11.1 *Left* Multi-layer component with joints. *Right* Heat flow through a real contact joint

Heat flow at the static contacting joint is displayed on the right. As shown, the actual surface of two components within contact (real surface area) equates to a fraction of the nominal surface area. This is due to material roughness, which occurs subject to the manufacturing process of the components surface. As a result, thermal resistance arises, which, in turn, causes temperature-difference between the two components at the joint. For the development of reliable models, precise knowledge of the heat transfer mechanisms at these joints is required. The aim of the research project is therefore to quantitatively describe the heat transfer mechanisms at the joints.

The relationship between the temperature drop at the joint and the heat flow through the contact surface is expressed by following Eq. (11.1)

$$\dot{Q} = h_c \cdot A \cdot \Delta T_c \tag{11.1}$$

whereby A represents the nominal area and ΔT_c the temperature difference between the component surfaces within contact. The parameter h_c describes the heat transfer coefficient, which links the temperature differences with the respective heat flow. Various heat transfer mechanisms occur at the contact area. These include thermal conduction through the solid bodies, as well as heat transfer through the fluid within the cavities. Radiation heat also has to be taken into account at high surface temperatures. However, thermal conduction through the joints constitutes the major part of thermal transmission (Yovanovich 2005), which therefore is focused on in detail below.

The relevant influencing parameters on thermal conduction are related to the consistency of the contact region and can be subdivided into three groups (see Fig. 11.2). On the one hand, the surface geometry of the non-deformed joint partners has significance and can be described by the parameters of roughness and slope. The

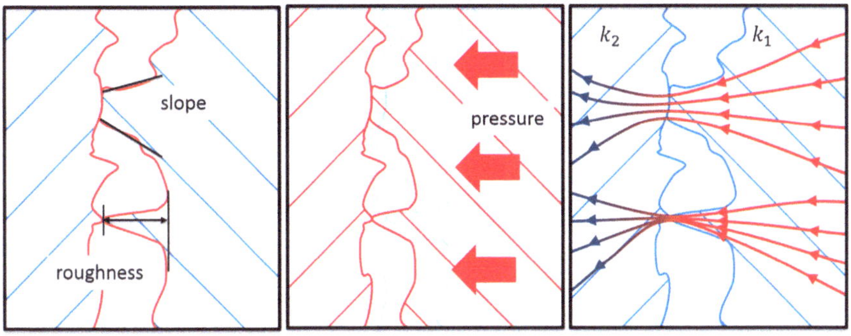

Fig. 11.2 Variables that influence contact transmission: surface structure (*left*), change in contact areas as a function of contact pressure (*middle*), thermophysical properties (*right*)

real contact area is also influenced by the nominal pressure on the joint (mechanical force upon the nominal surface area), since the contact ratio is changed due to deformation as a function of the mechanical properties. In addition to this, calculation of the heat transfer coefficient also requires the consideration of the thermophysical properties, such as the heat conductivity of both joint partners.

Within literature, various approaches are provided to describe the contact heat transfer coefficient h_c. One of the best known methods originates from Mikic (1974):

$$h_c = 1.13\tilde{k}\frac{\tan\theta}{R_q}\left(\frac{p}{p+H}\right)^{0.94} \tag{11.2}$$

The parameters in Eq. 11.2 include the harmonic mean of both components' heat conductivities $\tilde{k}$, the root mean square roughness R_q, the mean-angle of the roughness peaks (θ), the nominal pressure (p) and the material hardness (H). The correlation however, can only be used as an approximation, since the determination of the mean angle of the roughness peaks in particular, is unknown in most cases and its measurement takes considerable time and expense. The model is also based on the assumption that both specimen surfaces have Gaussian, isotropic height distribution with semi-spherical roughness peaks. It is assumed that the individual contact spots do not influence each other thermally or mechanically. This and other analytical approaches have already been refined and expanded by adding influencing parameters. Here, first of all, the model by Bahrami et al. (2004) should be mentioned, which takes into account heat transport mechanisms throughout the gas inside the cavities.

Figure 11.3 outlines a comparison of the heat transfer coefficients determined by means of the given Mikic model and the model by Bahrami et al. (2004) with experimental studies. Obvious deviations are shown both between the two models and in comparison with the experimental outcomes. As a result, this elucidates the open issues that have to be clarified by research in order to derive a reliable heat transfer model.

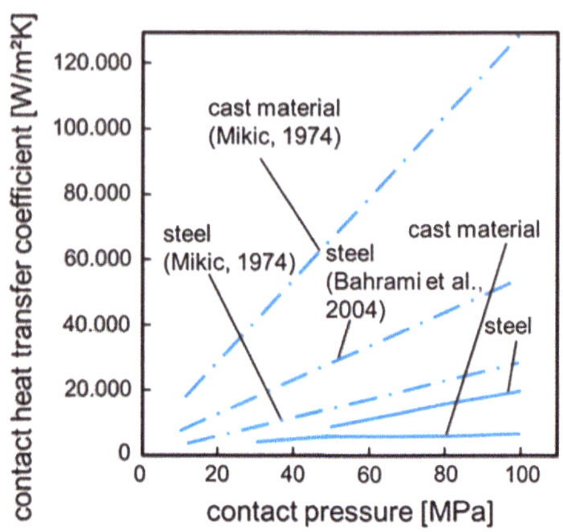

Fig. 11.3 Experimental contact heat transfer coefficients (*continuous lines*), compared with correlations by (according to) Mikic and Bahrami et al. (*dot-dashed-lines*) (Fieberg 2007)

11.2 Approach

11.2.1 Determination of Contact Heat Transfer in Experiments

To quantify the contact heat transfer by means of measurement, different requirements have to be fulfilled by the experimental setup and procedure. On the one hand, representative joints for machine tools need to be modelled and characterised as accurately as possible in terms of material and structural consistency. On the other hand contact pressure between the surfaces needs to be varied in a range that is relevant for the component joints. Since the goal of the experiment is to quantify heat flows (local temperature gradients) by means of temperature measurements, which, in turn, also permits conclusions regarding the heat transfer coefficient, stringent requirements have to be fulfilled by the measurement method's accuracy.

This is achieved by means of a high resolution infrared thermographic camera. The measurements are executed by using two specimens of similar surface and material properties as relevant machine components. As with the machine components under consideration, the specimens are pressed together at their faces within a hydraulic test machine. A photo of the test setup and a characteristic infrared image is shown in Fig. 11.4.

The investigations described here focused on the mild steel C45 (material no. 1.0503), which is, for instance, employed as machine tool tables and clamping fixtures.

These temporally resolved temperature data are used to determine the contact heat transfer. To do this, the test concept illustrated in Fig. 11.5 is implemented for the experimental studies.

Unequal initial temperatures cause heat transfer from the hotter to the colder specimen as soon as the two are brought into contact. The pressure-time curve is measured in addition to temperature distribution (Fig. 11.5, *bottom left*). With the

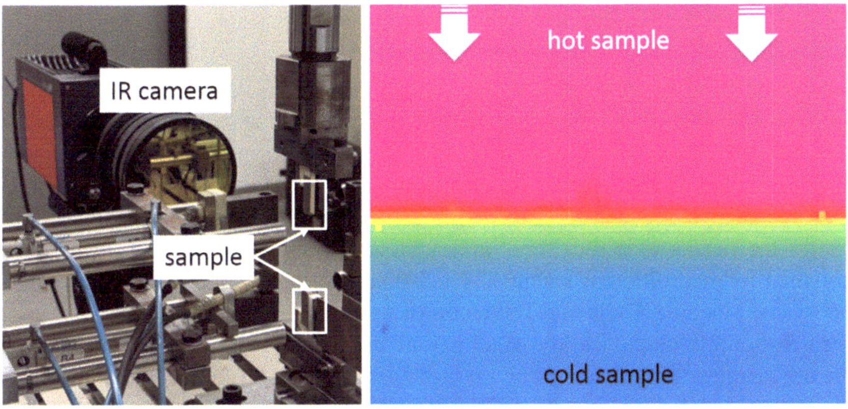

Fig. 11.4 Test setup (*left*); Infrared image (*right*)

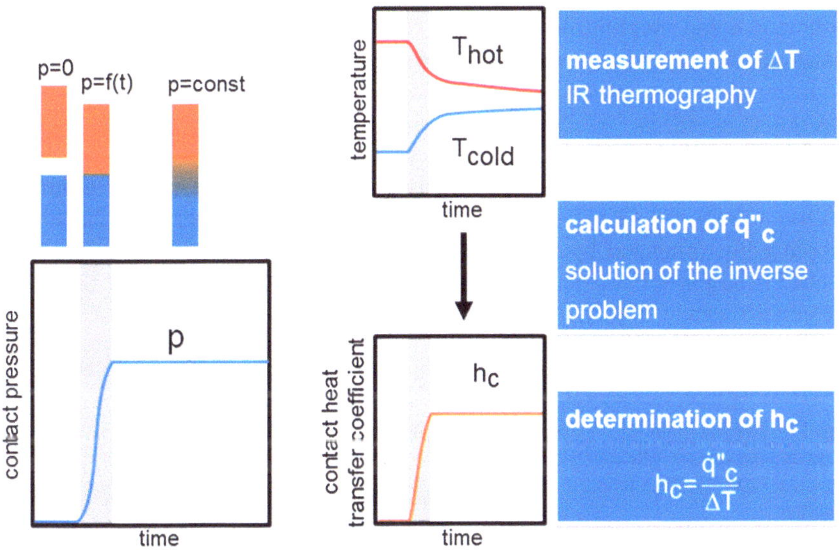

Fig. 11.5 Strategy and evaluation of the compression tests

specimens' temperature difference at the contact area, the solution to the respective inverse heat conduction problem is obtained in accordance with Fieberg (2007). Its solution resolves the value of the contact heat transfer coefficient (Fig. 11.5, *right*). Experimentally gained heat transfer coefficients not only make a direct quantitative prediction regarding heat transfer possible, but can also be used as validation data for numerical calculations.

11.2.2 Modelling the Contact Heat Transfer

Analytical correlations to quantify the contact heat transfer use several assumptions and simplifications, both regarding surface geometry and in terms of mechanics and heat transfer. The main assumptions reduce the surface to an isotropic structure with a Gaussian roughness profile. By this, the number of free parameters in surface representation is reduced to two: roughness and mean slope. Figure 11.6 demonstrates the results of a roughness and contour measurement using a caliper arm, obtained for two different technically machined surfaces, supported by the Laboratory for Machine Tools and Production Engineering (WZL Aachen). As can be seen from these outcomes, the assumptions of Gaussian distribution and isotropy are not justified for real, technically relevant surfaces, and technically manufactured surfaces may have different geometries despite similar characteristic values (in this case roughness value Ra).

This shows the necessity to validate the analytical correlations using experimental measurements and to numerically analyse the different surfaces regarding their mechanical and thermal contact behaviour without relying on geometrical assumptions and simplifications. For this purpose, a simulation tool was developed, which models the contact behaviour of generated surfaces with isotropic and Gaussian roughness, as well as with topographically measured technical surfaces (Frekers et al. 2014).

This calculation is performed in two steps. First the elastic semi-sphere model by Bhushan (1998) is applied and enhanced using a plastic component as developed by Willner (2004) and Görke (2010). The deformed surfaces' thermal behaviour is subsequently computed by means of an implicit finite difference technique. The contact heat transfer coefficient results from the calculated temperature drop over the contact area.

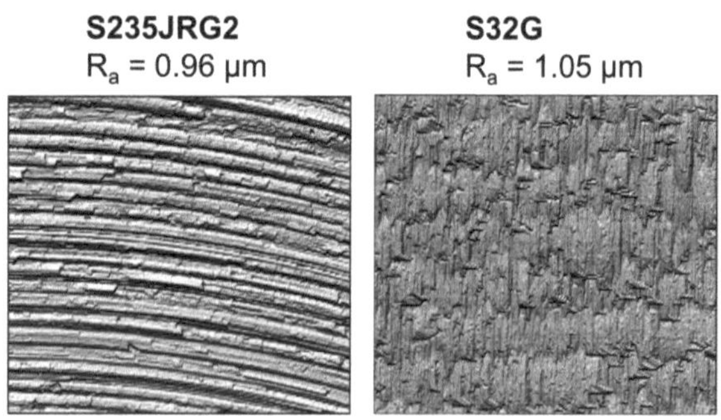

Fig. 11.6 Topographic measurement of two different technical surfaces with similar roughness

11.3 Results

The outcomes for the simulation of real and generated surfaces are taken from Frekers et al. (2014) and are compared to measurement results obtained from the experimental setup described. Contact heat transfer coefficients are computed and measured as a function of the nominal contact pressure. Mild steel C45 was used as the specimen material. The heat transfer coefficients elucidated in Figs. 11.7 and 11.8 are normalized by the conductivity of the material. This is done, since heat conductivity has a linear influence of heat transfer—both within the developed model and as stated by the Mikic-correlation. This makes the results more comparable for different joint materials.

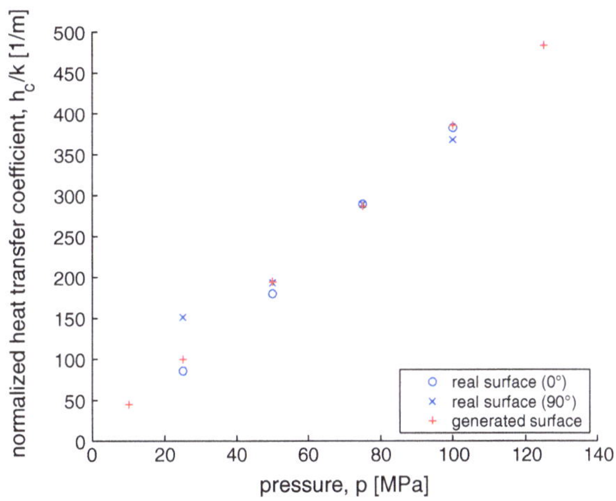

Fig. 11.7 Simulation outcomes of pressure-dependent heat transfer coefficients for real and generated surfaces (Frekers et al. 2014)

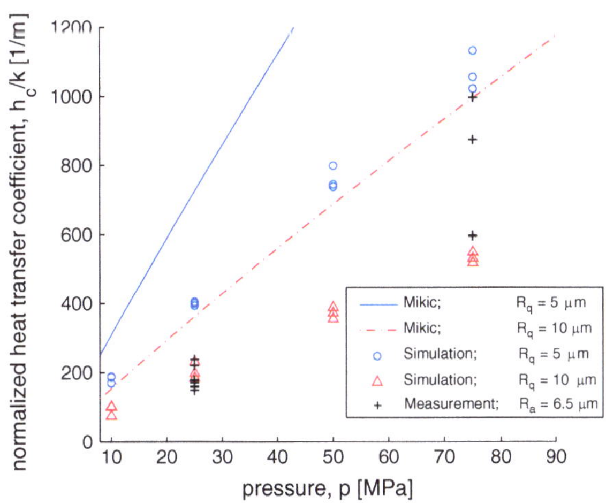

Fig. 11.8 Contact heat transfers—found in experiments and simulated, as well as found by approach correlations according to (by) Mikic

Figure 11.7 shows the comparison between a generated surface and a real measured technical surface. The real surface is simulated in contact with itself, first under a 0° angle (mirrored), and then rotated at an angle of 90° to one another. In comparison, the generated surface has identical roughness and mean slope.

The results demonstrate a change in the contact heat transfer at different contact angles, as well as differences between the transfer coefficients of real and generated surfaces. Overall however, the variation in contact heat transfer coefficient values between generated and real surfaces is low.

Simulation outcomes and predictions of the Mikic correlation are compared with experimental measurements in Fig. 11.8.

The shown experimental data each are obtained from different test specimens. The exact microscopic surface geometry of each specimen examined is, as is the case of problems commonly appearing in practice, not completely known. Only the mean roughness value (Ra value) was measured, to 6.5 μm. The different simulated surfaces have squared roughness values (Rq value) that amount to 5 or 10 μm, and a mean slope ($\tan \theta$) of 0.2 mm.

As the measured results indicate, the developed simulation model is able to predict contact heat transfer coefficients in the observed pressure range considerably better than the Mikic correlation. The measurement results strongly vary at high coefficients. A possible explanation for this is the measurement accuracy, since high heat transfer coefficients make the analysis procedure more fault-prone. It should also be noted that the surface geometry of the experimental specimens is only insufficiently represented beyond the roughness value. Slightly differing contact geometries also result in greater scattering of outcomes in cases of increasing heat transfer coefficients. From these observations and findings, future research needs can be identified clearly.

11.4 Classification of Outcomes in the CRC/TR 96

The model developed and the experiments performed enable the quantification of the heat transfer coefficients at component joints. By using these experimental and numerical methods, the parameters can be derived for each joint and integrated into the corresponding calculation tool. For many components within a machine tool, the thermal resistance at the joint is crucially relevant for the thermal behaviour of the whole machine and thus, decisive for model accuracy. Hence, the heat transfer coefficients quantified here make precise representation and modelling of the machine tool possible.

11.5 Outlook

The described investigations deal with the experimental exploration of C45 as specimen material. Thus, continuing research is aimed to using this methodical approach to characterise all relevant steel types employed in machine tool building.

Nevertheless, reliable models for an exact representation of the heat transfer behaviour as a function of geometric and structural mechanical parameters that can be measured easily are still needed. It is particularly important to provide data for a wide scope of varying pressures and geometry values. For this reason, ongoing research activities are aimed at the steady refinement and adaptation of the simulation models.

References

Bahrami M, Culham JR, Yovanovich MM, Schneider GE (2004) Thermal joint resistance of conforming rough surfaces with gas-filled gaps. J Thermophys Heat Transf 18:318–325

Bhushan B (1998) Contact mechanics of rough surfaces in tribology: multiple asperity contact. Tribol Lett 4:1–35

Fieberg C (2007) Kontaktwärbergang unter hohen Druck- und Temperaturbedingungen. Dissertation, RWTH Aachen

Frekers Y, Vieler S, van Buren S, et al (2014) Mechanical simulation. In: 15th International heat transfer conference (accepted)

Görke D (2010) Experimentelle und numerische Untersuchungen des normal- und Tangentialkontaktverhaltens rauer metallischer Oberflächen. PhD. thesis, Universität Erlangen

Mikic BB (1974) Thermal contact conductance; theoretical considerations. Int J Heat Mass Transf 17:205–214

Willner K (2004) Elasto-plastic normal contact of three-dimensional fractal surfaces using halfspace theory. J Tribol 126:28–33. doi:10.1115/1.1631019

Yovanovich MM (2005) Four decades of research on thermal contact, gap, and joint resistance in microelectronics. IEEE Trans Compon Packag Technol 28:182–206. doi:10.1109/TCAPT.2005.848483

Chapter 12
Investigation of Components and Assembly Groups

Christian Brecher, Marcel Fey, Dorothea Haber and Kolja Bakarinow

Abstract Machine tools comprise thermally active and passive components. This results in a heterogeneous temperature field, which may cause displacements inside the machine and, ultimately, a TCP displacement. In subproject B03, the individual components are subjected to metrological study and transformed into a simulation model in order to obtain a simulation representation of the thermo-elastic machine tool behaviour. In the methodology to determine the complete machine behaviour, a distinction is made between a top-down- and a bottom-up approach. In the first case, the behaviour of the complete machine is analysed and synchronized with the simulation model. Thus, for instance, it is possible to represent the behaviour of the passive components. In the second case, the thermally active components—ball screws, guidance systems and the main spindle—of the machine tool are individually studied on test beds. Information gleaned from the investigations conducted on the test beds and the models derived from them are added to the complete machine model, where they are employed to identify the complete temperature- and displacement field. To exclude the ambient influence as much as possible, the test beds are installed in a climatic chamber.

C. Brecher (✉) · M. Fey · D. Haber · K. Bakarinow
Laboratory for Machine Tools and Production Engineering (WZL),
RWTH Aachen University, Aache, Germany
e-mail: c.brecher@wzl.rwth-aachen.de

M. Fey
e-mail: m.fey@wzl.rwth-aachen.de

D. Haber
e-mail: d.haber@wzl.rwth-aachen.de

K. Bakarinow
e-mail: k.bakarinow@wzl.rwth-aachen.de

K. Großmann (ed.), *Thermo-energetic Design of Machine Tools*,
Lecture Notes in Production Engineering, DOI 10.1007/978-3-319-12625-8_12

12.1 Introduction

In subproject B03 on the "Investigation of Components and Assembly Groups", the aim is to extend the state of knowledge of the thermal behaviour of machine tool components and subassemblies, as well as of machine tools, depending on different thermal stress states and intensities. In this process, the components can be subdivided into the following two subgroups:

Firstly, a machine tool is made of thermally active components, generating heat by friction and electrical losses and, in this way, causing that heat to enter into the machine structure. Tactile temperature sensors are used for measurement, on the one hand. On the other hand, newer developments indicate (Koscàk 2007) that investigations by means of thermographic cameras make sense in order to make a surface-related statement about the temperature field. This approach is pursued in detail in the studies. In the CRC/TR 96, the following components are explored in more detail: main spindle, ball screws (Heisel et al. 2005; Kunc 2013; Chang 2009; Ess 2012) and linear guideways (Kunc 2013; Ess 2012). Drives that also introduce heat into the system are explored, with the results that are obtained being made available to the subproject C04 (see Chap. 19). Investigation is here focussed on a targeted variation of influencing parameters, such as revolutions per minute and feed. Secondly, a machine tool also consists of passive components. They absorb heat and transfer it to the adjacent components, such as, for instance, the machine bed (Neugebauer 2001), machine table and the slides, or dissipate it into the environment.

Due to the variety of materials these machine tool components are made of, as well as to the heterogeneous distribution of the heat sources and heterogeneous heat transitions, an unpredictable inhomogeneous temperature field is built up in the machine tool (Brecher et al. 2011; Mayr 2009; Neugebauer et al. 2002), which is extremely difficult to define because of the unknown constraints. To determine the thermo-elastic deformations by means of a simulation model (Neugebauer et al. 2001, 2002), one has to know the heat sources and sinks, as well as the heat transfer conditions, to define the emerging temperature field and the resulting displacements. The parameterisation of a model like this is conducted not only through analysis of the individual components, but also by investigation of the complete machine. The individual components are explored in terms of their thermo-elastic behaviour on test beds, while the complete machine is loaded as a whole. In this process, both the temperatures in the complete structure and the displacement at the Tool Center Point (TCP) are determined. In the next step, models of the individual components are generated, which are kept as simple as possible, but nevertheless map the thermo-elastic component behaviour. These individual models are then integrated into the model of the complete machine. This overall model allows for mapping of various load cases and alignment with experiments carried out using the entire machine.

The final objective is to make use of the results and models from the subproject, both for control-internal correction and for compensation in machine tool design and dimensioning (Ess 2012; Heisel and Stehle 1996). By varying the settings, it is possible to create simulation models that can represent the thermo-elastic component behaviour very realistically and thus promise high predictive accuracy.

12.2 Approach

The objectives described above are achieved by the following two different approaches shown in principle in Fig. 12.1 (left). The first approach followed is a bottom-up one. The thermally active machine tool components are investigated individually on test beds. These components are ball screws, guidance systems and the main spindle. Information resulting from the test bed investigations and the models thereby derived are added to the complete machine model and are employed in the model. To achieve high accuracy in the individual models, the same components that are employed in the complete machine, are tested in the test bed investigations. The machine tool is depicted as a CAD model in Fig. 12.1 (right).

The second approach traced is the top-down strategy. This means that the behaviour of the complete machine tool is analysed and is also aligned with the simulation model. This makes it possible, for example, to map the thermo-elastic behaviour of the passive components. Since the heat transfer conditions in the machine are different from those on the test beds, additional investigations of the complete machine tool are necessary. Continuous alignment of the two approaches makes it possible to provide the optimal simulation result.

12.2.1 Guidance Systems

The test bed to investigate guidance systems is shown in a schematic view in Fig. 12.2 (left). Two guide rails of the same structure are interlocked via load unit.

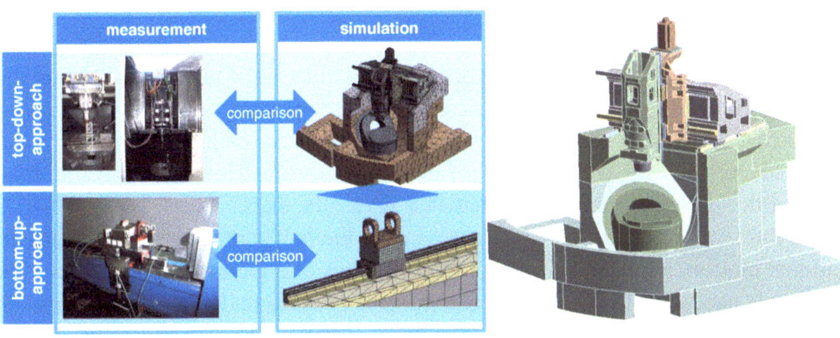

Fig. 12.1 Top-down and bottom-up approaches (*left*), Simplified CAD model of the complete machine (*right*)

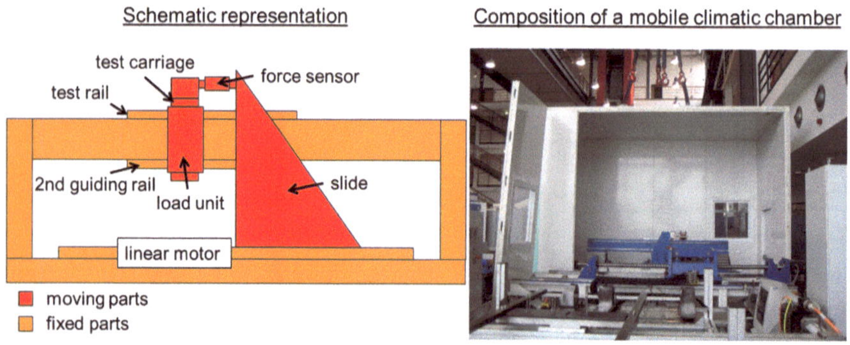

Fig. 12.2 Test bed for guidance systems

Both guide shoes are driven by a linear motor. The linear drive is linked with the guide shoe by a slide and is thus locally and thermally discoupled. A sensor for frictional force measurement is situated between the slide and the moving parts. The frictional force is crucial to quantify the thermal output entering the system.

In the experiment, the guide shoes were traversed at constant speed under a constant fastening force. The traversing motion stops every 2 min, and the temperatures are measured both by thermocouples and a thermographic camera. Holes cut into the upper test guide shoe and the guide rail make it possible to measure the temperatures inside at different positions. The temperature curves on the surface were recorded by means of a thermographic camera, and PT100 sensors were attached for alignment. The frictional force is also measured every 30 min. In this way it was possible to trace the change of the frictional force depending on the temperature trend.

Variations in the ambient temperature do not affect the investigations, since the test bed, as represented in Fig. 12.2 (right), is housed in a mobile climatic chamber.

12.2.2 Ball Screws

The ball screw for the Z axis of the complete machine tool was investigated on the test bed shown in Fig. 12.3. The ball screw is driven by a servomotor by means of a dog clutch. The nut is connected with the slide via a load cell, at which both axial force and torsion are measured. A linear motor, transferring a specific load opposing the traversing motion, is installed under the slide.

To be able to quantify the temperature inside the ball screw despite the constant rotational motion, temperature is recorded by means of a thermographic camera. To do this, it is necessary to blacken the components in order to determine the emission coefficient. Furthermore, reflections from the environment on shiny metal cause a falsification of measurement results. PT100 sensors are attached to the structure for alignment and thus quantify the emission coefficient.

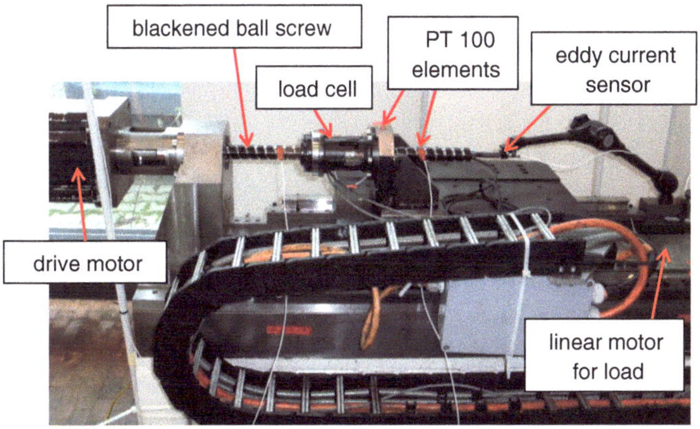

Fig. 12.3 Test bed for ball screw investigation

As was also the case in the experiments with the guidance systems, these are continuous load tests. The ball screw is driven at a constant speed over 8 h and subjected to a constant linear drive load until a steady state has been achieved. The test bed then stands still and the components cool down. Since the warming up and the cooling down phases have different performance characteristics, temperatures are also measured in the cooling down stage.

To exclude the impact of the ambient temperature to the greatest extent possible, the test bed is shielded from sunlight and draught, and ambient temperature is recorded for monitoring.

An eddy-current sensor is applied on the spindle's front face and thus traces spindle lengthening as a result of warming up. This is not only relevant for modeling, but also for a correction of displacements later on.

12.2.3 Demonstrator Machine Tool

Both approaches are investigated on the demonstrator machine. Experiments are conducted to apply the top-down approach. The measuring setup used is illustrated in Fig. 12.4 (left). A test mandrel made of Invar steel is fastened to the machine tool interface, and a measuring angle is mounted on the machine table. In this angle, there are fixed—as recommended in ISO 230-3—five path sensors to capture the TCP displacements in X-, Y- and Z directions, as well as inclination related to the X- and Y axes. Heat is introduced into the machine tool by traversing a linear axis. At first, each axis was traversed individually in the air cut to consider the heat sources separately. The subject of investigation is a five-axis machine tool whose Z axis is driven via a ball screw, and the remaining axes are driven by linear drives. Apart from the TCP displacements, temperatures and strains were metered in the machine tool structure for alignment with the simulation model, see Fig. 12.4 (right).

Fig. 12.4 Measurement setup for the entire machine tool (*left*), meshed model (*right*)

12.3 Results

12.3.1 Rail Guidance Systems

The frictional force emerging during the traversing motion was captured on the test bed introduced in Fig. 12.2 by means of a piezoelectric force sensor at 30-min intervals. The average frictional force is shown for various traversing speed values as a function of the measurement quantity in Fig. 12.5. It is clear that frictional force increases as a function of speed. However, this response seems to be non-linear, as expected. The zero position of the frictional force shown at the beginning and at the end can be explained by the fact that the test bed stands still at the moment of measurement, and thus no friction emerges. From the diagram, it can also be abstracted that the frictional force is constant over the duration of the test. As a consequence, the temperature in the guidance system does not affect, or only

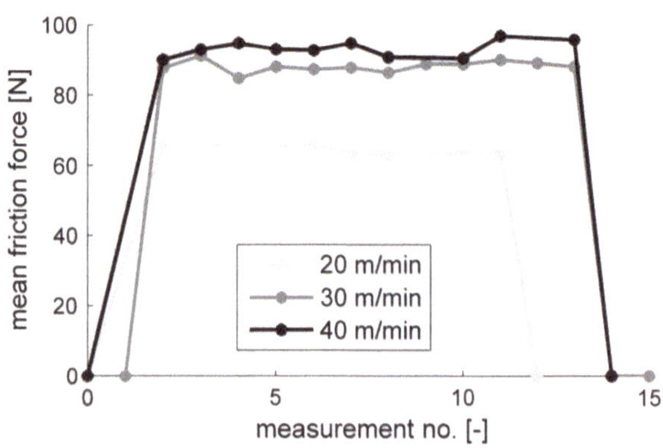

Fig. 12.5 Average frictional force for various speed values as a function of the test duration

negligibly affects the frictional force and, in turn, the thermal dissipation output. The friction power is calculated using the frictional force and the speed, which can be read from the control. This power is almost completely transferred into thermal output, so that it can be used as an input variable in the models. The thermo-elastic behaviour of a machine component can be mapped by simulation by using the power produced by friction. In the test, an adjustment with the component temperatures measured is executed for a view that very closely approximates reality. This partial model shapes the basis for the entire simulation model of the machine tool.

12.3.2 Ball Screws

The explorations of the ball screw for the Z axis of the entire machine were conducted for several speeds and load values to characterise the component's holistic behaviour. As previously mentioned, the warming up phase lasted approximately 8 h and was followed by a cooling-down stage, in which the test bed does not traverse any more. This made it possible to attach PT100 sensors to the spindle in this phase to align them directly with the thermographic camera images and to calibrate the thermographic camera. For alignment, additional temperature sensors were attached to the back of the load cell and the nut. In Fig. 12.6, the emission coefficient is still one at each point.

For systematic evaluation of the ball screw's thermal behaviour, the temperature curve was first analysed at individual points. Figure 12.7 represents the exemplary

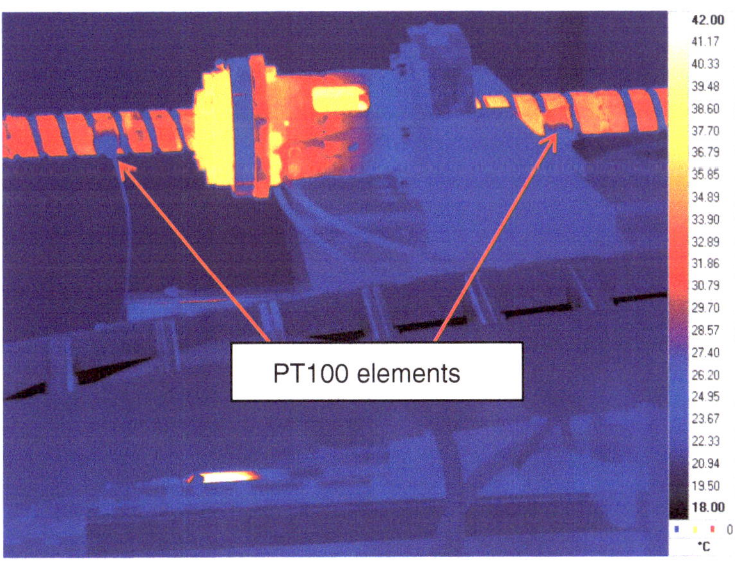

Fig. 12.6 Thermographic representation of the ball screw test bed

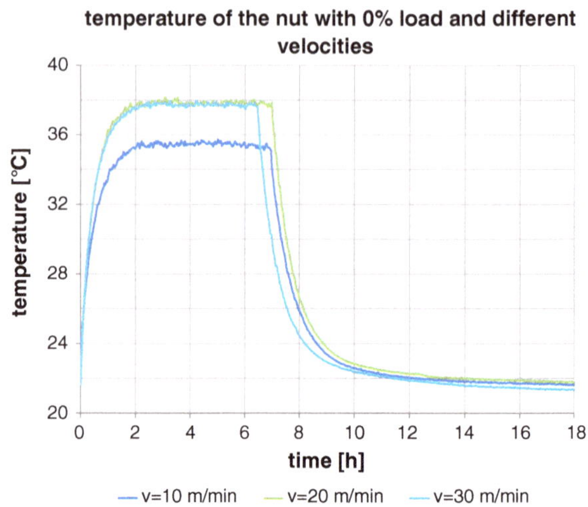

Fig. 12.7 Temperature of the ball screw nut at several speeds

temperature curve for the ball screw nut at several speed values, though without the linear drive's load.

The characteristic exponential performance both in the warming up- and the cooling down phases can be seen. At the minimal speed of 10 m/min, the maximal temperature of approximately 35 °C appears, and, with the highest speed value of 30 m/min, the maximal temperature is 38 °C. The steady state was established after approximately 2.5 h, and the temperature remained constant after this time. Due to the fact that the nut has almost the same temperature at a speed of 20 m/min and at the higher speed value, it is still impossible to predict the precise relationships. Consequently, additional measurements and an ongoing analysis of the measurements already carried out are required for a complete characterisation. In this process, the thermographic camera photos after the determination of the specific emission coefficients play an important part in the analysis of the heat flows and temperature fields.

12.3.3 The Machine Tool as a Whole

The investigations described above became a part of the simulation model of the complete machine tool in the bottom-up approach. For individual load cases, the heat input from the guidance systems and the ball screw were introduced into the complete machine tool. Temperatures were also measured for the structural parts and the drives in the measurements of the machine (top-down approach). They were employed to align the component investigations. However, they also bridged the gaps in components which could not be modelled until adequate models had been stored for them. Thus, a temperature distribution could be calculated in the machine tool simulation, which conforms relatively closely to the measurements (Fig. 12.8).

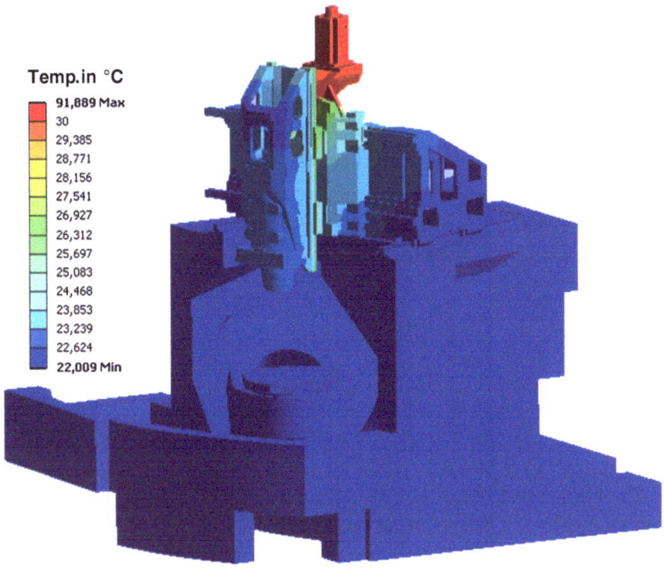

Fig. 12.8 Simulated temperature distribution for the complete machine tool

In this case, the Z axis is traversed, whereas the X- and Y axes are not traversed, so that the heat introduced into the guide rail and ball screw of the Z axis is limited. To represent the thermo-elastic behaviour reliably and also for various load cases, however, it is necessary to store both more precise models for the components and to parameterise also other constraints as well, such as convection and heat transitions between components. Thus, for instance, the temperature shown for the Z slide is still too low, since the heat build-up of the spindle controlled in closed-loop has not yet been modelled.

12.4 Classification of Outcomes CRC/TR 96

A thermally suitable design for machine tools resulting in a reduction in thermally induced machine tool errors demands exact knowledge of the thermo-elastic behaviour of the machine components and subassemblies, and, as a result, of the complete machine tool. This behaviour is explored and subjected to a systematic consideration in the subproject B03 with the aim of working in conjunction with the other subprojects, such as A05, B02, B05 and C04, to find a solution to the conflict of interests arising in the overall CRC/TR 96 among energy use, accuracy and productivity.

12.5 Outlook

The studies demonstrate that it is already possible to determine the temperature field that appears in the machine as a whole for individual load cases with limited accuracy. However, optimisation measures are necessary for the further use of the model. In this process, the constraints, such as convection, heat transitions, and the response of individual components under different loads have to be determined, and, finally, the temperature fields have to be transferred to the displacement of the machine.

References

Brecher C, Wissmann A, Klein W (2011) Thermisch bedingtes Verformungsverhalten von Werkzeugmaschinen. ZWF Zeitschrift für wirtschaftlichen Fabrikbetrieb 106:663–666

Chang C-F (2009) A theory of ball-screw thermal compensation. In: International multiconference of engineers and computer scientists

Ess M (2012) Simulation and compensation of thermal errors of machine tools. Dissertation, ETH Zürich

Heisel U, Stehle T (1996) Ermittlung und Kompensation thermisch bedingter Deformationen an Werkzeugmaschinen und Industrierobotern. Produktion und Management. Produktion und Management 86

Heisel U, Stehle T, Koscàk G (2005) Thermisches Verhalten von Kugelgewindetrieben. wt Werkstattstechnik 320

Koscàk G (2007) Ermittlung des instationären thermischen Verhaltens von Vorschubachsen mit Kugelgewindetrieb mit Hilfe der Verarbeitung thermografischer Messdaten. Dissertation, Universität Stuttgart

Kunc M (2013) Identifikation und Modellierung von nichtlinearen Dämpfungseffekten in Vorschubachsen für Werkzeugmaschinen. Dissertation, RWTH Aachen

Mayr J (2009) Beurteilung und Kompensation des Temperaturgangs von Werkzeugmaschinen. Dissertation, ETH Zürich

Neugebauer R, Harzbecker C, Nestmann S (2001) Optimization of thermal behavior of machine tools with separated machine bed components. Prod Eng Res Dev Germany Ann German Acad Soc Prod Eng 8:137–140

Neugebauer R, Weidermann F, Nestmann S (2002) A modeling approach to optimize the thermal behaviour of machine tool components. Production Engineering vol IX/1. Production engineering. Research and development

Chapter 13
Adjustment of Uncertain Parameters in Thermal Models of Machine Tools

Bernd Kauschinger, Klaus Kabitzsch and Steffen Schroeder

Abstract Thermal models of machine tools contain parameters that represent machine-specific and time-variable properties. In the design process, these parameters cannot be estimated with sufficient accuracy. Thus, they have to be adjusted by measurements. At present, substantial time, effort and expensive measurement equipment are required for adjustment, as well as in-depth expertise. Consequently, the goal is to develop cost efficient methods for rapid and comprehensive adjustment. This is to be achieved using a systematic strategy for the support and automation of adjustment processes. The strategy is demonstrated based on a thermal model of a bearing assembly.

13.1 Introduction

In the scope of the CRC/TR 96 measures for passive compensation and active correction of thermally induced deformations as well as for a thermo-energetic optimized design of machine tools are being developed. Basis of these measures are thermal behaviour models of the machine and its assemblies. They are used as a tool for evaluation and optimisation of relevant design features as well as for the generation of correction values for the feed axes. In the latter case there are particularly high demands on model accuracy.

B. Kauschinger · S. Schroeder (✉)
Faculty of Mechanical Engineering, Institute for Machine Tools and Control Engineering, Technical University Dresden, Dresden, Germany
e-mail: steffen.schroeder1@tu-dresden.de

B. Kauschinger
e-mail: bernd.kauschinger@tu-dresden.de

K. Kabitzsch
Faculty of Computer Science, Institute for Technical Information Systems, Technical University Dresden, Dresden, Germany
e-mail: klaus.kabitzsch@tu-dresden.de

K. Großmann (ed.), *Thermo-energetic Design of Machine Tools*,
Lecture Notes in Production Engineering, DOI 10.1007/978-3-319-12625-8_13

Thermal models can be distinguished in three categories: correlative, property-based and structure-based models (Großmann 2012). Correlative and property-based thermal models are based on empirical model approaches. Hence, the validity of these models is limited in general. All model parameters need to be adjusted by means of experimental measurement on the real machine. These models are only suitable for correction purposes. Structure-based thermal models use physically founded approaches. This results in a principally wider validity. These models are suitable not only for correction (Chap. 16) but also for analysis and optimization in the design process or compensation measures (Chap. 17). Therefore, subproject B04 focuses on structure-based models.

In contrast to correlative and property-based models most of the parameters in structure-based models can be estimated from design data with sufficient accuracy. Hence there are only a few parameters that need to be adjusted based on experimental measurements. But the adjustment procedure and suitable experiments need to be designed specifically for every machine (resp. model) and expert knowledge of the thermal behaviour is required for that. Up to now the design and execution of the adjustment procedure is very inefficient, since there are no (software) tools for support and automation available that encapsulate the required expert knowledge. The need for research resulting from this is addressed in subproject B04.

13.1.1 Uncertain Parameters in Thermal Models

Structure-based thermal models describe the whole thermal causal chain. Model inputs are technological data generally available in the control, such as spindle speed or feed rate. They are used to calculate thermal loads (power loss, heat flow). Based on these loads the transient temperature field and the resulting thermal deformation of the machine's assemblies are calculated. By applying the position-dependent kinematic transformation finally the thermally induced displacement of the tool center point (TCP) is determined (Großmann and Jungnickel 2001).

Uncertain parameters that need experimental adjustment are located in model parts for the calculation of thermal loads and temperature fields. Notably these are *convective heat exchange* and *generation of power loss*. Convective heat exchange takes place between the machine and the environment as well as in fluid circulation systems. Power loss is caused by electrical processes in drives and by friction in guiding elements.

Rolling bearings are a typical example of these guiding elements. Despite their low friction coefficient in conjunction with the usual speed range, rolling bearings generate noticeable power losses and can thus become the main heat sources in the machine. This is especially the case in high-speed main spindles. In addition, the heat of the inner and outer ring of the bearing is exchanged via the rolling elements. As a result, rolling bearings substantially affect the temperature distribution in the machine and thus also thermal deformations relevant for accuracy (Großmann 1998).

Power loss of rolling bearings is generated by friction. The major proportion of heat arises in the moving contact between the rolling elements and the bearing rings. Friction is influenced by several coupled elasto-hydrodynamic processes (Großmann and Jungnickel 2003). Essential processes that appear are:

- Transition from Coulomb friction to viscous friction as a nonlinear function of speed,
- Sliding- and rolling-off procedures in the contact,
- Elastic deformation of the solids in contact caused by preload,
- Friction generates power loss and therefore changes the bearing temperature,
- Change in preload due to thermally induced deformations of the bearing assembly and
- Change in lubricant viscosity caused by temperature and speed.

These phenomena result in a frictional torque of the bearing, which depends on many influences and shows explicitly nonlinear and time-variable characteristics depending on speed and temperature (Fig. 13.1). This frictional behaviour has been studied in a comprehensive manner previously (Bäumler et al. 2012). As a result there is a variety of friction model approaches that can be used as submodels in the structure-based thermal model. Due to the complex influences commonly used model approaches are mainly of empiric nature. Though they are able to reproduce relevant effects sufficiently, there are significant uncertainties in the model parameters, resulting in substantial model errors. The investigations by Fritz (2011) should be mentioned as an example. Here calculated friction coefficients differed from measured ones by 30–200%.

The approaches to determine the parameters of friction models are one reason for the uncertainties (Jedrzejewski et al. 1989). In some models, for instance, the average diameter is taken out of many geometric values to determine a parameter that represents the geometric influence. The exemplary nature of the influences of friction due to manufacturing and assembly tolerances is another reason for

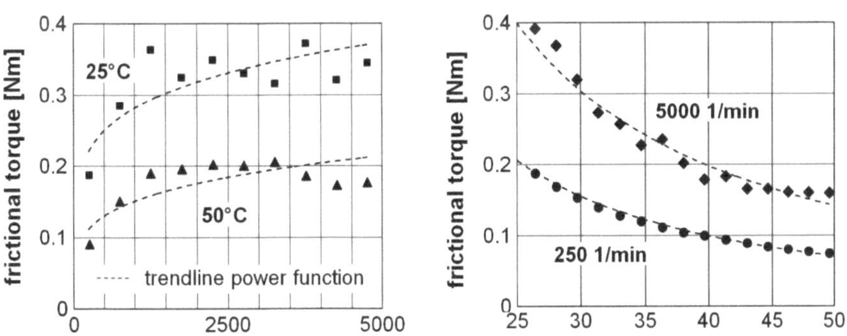

Fig. 13.1 Measured temperature- and speed-dependent bearing friction of an angular contact ball bearing of the 7207-B-2RS type

parameter uncertainty, e.g. bearing preload. Even close geometric tolerances of the bearing assembly components may lead to considerable deviations in preload. Last but not least, another reason can also be found in the bearings' running-in behaviour, which is characterised by changing frictional conditions. Measurements by Nachimowicz and Jałbrzykowski (2012) have shown a long-term decrease of the frictional torque of angular contact ball bearings by 60 %.

13.1.2 Adjustment of Uncertain Parameters

In general, total model uncertainty results from the selected model approach (model structure) and its parameters. The parameter uncertainty can be machine-related classified in type-identical or instance-related uncertainty. In the case of type-identical uncertainty, the parameter values only slightly scatter among several machines of the same type, and, consequently, an adjustment is only needed once for each machine type. However, in the case of instance-related uncertainty, the properties scatter widely, e.g. due to manufacturing and assembly tolerances. In these cases, the parameters have to be adjusted for each individual machine instance. Parameter uncertainty can also be distinguished in terms of the long-term characteristics of the real properties. If the properties significantly change over time due to running-in and wear effects, then it is necessary to repeatedly readjust the parameters. A summary of parameter uncertainty classification under parameter adjustment aspects is given in Table 13.1. Most parameters that need an adjustment are instance-related.

In addition to the specific background of the uncertain parameters mentioned above, there are essential boundary conditions that have to be regarded in parameter adjustment of thermal models:

- Almost the entire machine structure as well as the surroundings are relevant for behaviour. Additionally several physical areas (thermal loads, temperature fields, deformations, kinematics) have to be represented along the thermal causal chain, which demands comprehensive and complex models.
- Many properties emerge during assembly and use of the machine. Hence an adjustment can only be conducted on the assembled machine where access for metrological purposes is strongly limited and only indirect measurements are possible.

Table 13.1 Classification of parameter uncertainties

	Machine related		Long-term related	
Parameter uncertainty	Type-identical	Instance-related	Constant	Variable
Parameter adjustment	Once for machine type	For every machine instance	Once	Several times

- Submodels with uncertain parameters typically include nonlinear approaches that complicate the adjustment.
- The thermal behaviour is transient and its time constants range from minutes to hours. Consequently the parameter effect is substantially delayed and measurement times are long.
- Because of strong thermal coupling of machine parts individual parameters show far-ranging effects. Thus, the effects of different parameters overlap.
- Initial temperature values as a result of thermal history have to be predefined for the thermal model. This requires a special methodology, since it is impossible to measure the whole temperature field in an economical way.
- The adjustment process demands several sub-steps, such as analysis, planning, measurement and mathematical optimization. This requires widely varying approaches. These sub-steps have to be linked, which, in turn, means a high number of information-processing subtasks.

The performance of previous parameter adjustments of thermal machine models (Großmann and Jungnickel 2006) shows numerous shortcomings. Thus the tests can only be carried out under laboratory conditions. Only some dominant parameters of selected assemblies, such as bearing power loss of the main spindle, are adjusted. The load cases to stimulate the parameter effect are of very simple design. The parameters are successively investigated. Typically constant loads to stimulate one parameter at a time are applied. The resulting temperature and deformation step responses needed for adjustment take a long time. As a result performing all load cases can last from a few days to weeks. The parameter's effect is mostly traced by tactile measurements of the TCP displacement and additionally a few temperature sensors. Model and measurements are adjusted using manual empirical adaptation of parameters (by intuition) during the iterative comparison of simulation and measurement. Altogether, many manual operations and considerable amount of technical expertise are required for this methodology.

To eliminate the existing deficits, a method for rapid, comprehensive, low-cost and low-time adjustment of uncertain parameters has to be developed. This solution entails the engineering of efficient methods that permit support and automation of the adjustment. A fully automated implementation of the model under industrial conditions is the long-term objective.

13.2 Approach

13.2.1 Engineering of Efficient Parameter Adjustment Methods

The development of an efficient parameter adjustment procedure starts with making available supporting methods to solve the subtasks of adjustment. The boundary conditions are most of all predefined by the specific features of the thermal models

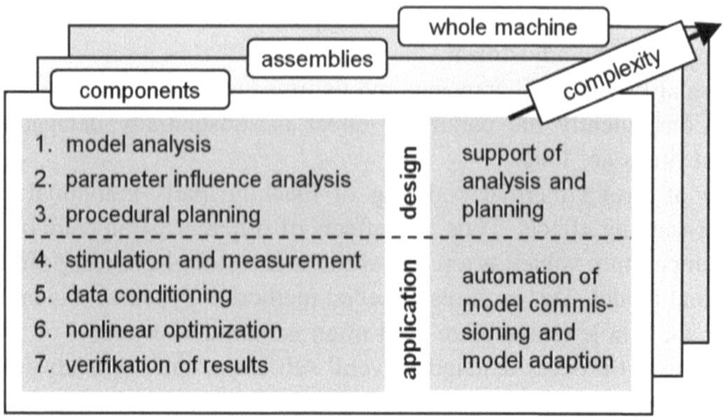

Fig. 13.2 Development strategy for efficient parameter adjustments

mentioned above and the measurement and adjusting opportunities on the machine tools. The studies assume that the model already exists, and thus do not intend any modelling. Another precondition is that the fundamental strategy for parameter identification and the mathematical methods required for nonlinear optimisation are known (e.g. Marquardt 1963; Davidon 1991). These methods are applied as state of the art and as they are. Rather, the objective is to design the comprehensive process of parameter adjustment to efficiently get appropriate data for the numerical optimization.

The schematic diagram of the development of an efficient parameter adjustment is shown in Fig. 13.2. The main tasks are split according to the engineering of methods applied to the individual procedural steps as well as to their formalisation and automation.

A procedural steps indicated in the Figure involve the following subtasks to be solved successively:

1. In a *model analysis* fundamental issues dealing with the uncertainty of parameters as well as their effect on the TCP displacement are investigated.
2. Ideal measuring points and stimulation for parameter adjustment are taken from the *parameter influence analysis*. The extent to which uncertain parameters affect state variables along the thermal causal chain and where these parameters affect the machine structure locally are investigated. Additionally the stimulation of parameters by load variables is explored.
3. *Procedural planning* provides the best suited and possibly least time-consuming load case for the stimulation of parameter effects and explores which data can really be acquired at given technical-economic and operational conditions. Aim of this step is an optimal design of the adjustment experiment.
4. The *stimulation and measurement* implies not only data acquisition at the real machine but also the setup and configuration of the experiment.

5. The *data conditioning* implies disturbance reduction (filtering, averaging), compression as well as transformation of the raw data acquired in step 4 to prepare it as input data for the optimization.
6. The *nonlinear optimization* step includes the extension of the thermal model for a simulation of the complete experiment, the selection and configuration of a mathematical optimization algorithm and the calculation to adjust model parameters itself.
7. Finally a *verification of results* is carried out, e.g. by means of characteristic indicators, such as residual errors and statistical data as well as evaluation measurements at the real machine.

These 7 steps are developed and investigated in terms of increasing model complexity. Starting from simple machine elements, such as bearings and ball screws, with a low number of uncertain parameters, complexity is increased via individual axis assemblies up to the complete machine tool. Hence the measurements will become more indirect and the measuring points are located farther from the optimal positions with the strongest parameter impacts. In the investigations, the number of measurements and the requirements to be fulfilled by the measurement equipment are reduced step by step to finally achieve a parameter adjustment procedure that appropriately works under real operational conditions of the machine tool.

13.2.2 Information Processing Methods for Parameter Adjustment

An efficient solution of the subtasks mentioned above needs support by software tools. As a basis for these tools customized information processing methods have to be engineered. A systematic and formal methodology is created and implemented in a demonstration software. From the analysis and planning tasks of the conceptual design stage (Fig. 13.2, steps 1–3) to parameter adjustment in the application phase (Fig. 13.2, steps 4–7) itself the software has to assist in subtasks with different content.

In addition to these generic concepts that are to be developed individual sub-steps can be improved using innovative techniques from other domains of computer science. Thus, the use of *Virtual Sensors* (Dementjev and Ribbecke 2011) would provide a solution for the acquisition of essential measurement variables that are important for the behaviour models but difficult to access by the available measurement equipment.

Furthermore, the techniques of automated design based on ontologies, like those presently in use in building automation, can be modified and applied to sub-steps of parameter identification (Dibowski 2013; Özlük 2013; Dibowski and Kabitzsch 2011; Ryssel et al. 2011; Dibowski et al. 2010). This would make it possible for software tools to generate recommendations for additional measurement equipment (Fig. 13.2, step 3). This is necessary, for instance, in cases in which there are no measurement devices available to capture relevant abstract measuring values of the behaviour model. Methods of automated design can also be adapted profitably to identify suitable conditioning steps (Kubin 2009) for the conditioning of measured values (Fig. 13.2, step 5) and to define their processing sequence.

13.3 Results

13.3.1 Visualisation for the Parameter Influence Analysis

The first parameter adjustment studies were conducted on bearing assemblies. A bearing test bed was engineered and built to be used for the explorations and the proof of efficient adjustment methods. The test bed consists of a pair of preloaded angular contact bearings and surrounding components relevant for function (Fig. 13.3). The surrounding parts are engineered to adjust the bearing loads as well as the bearing boundary conditions in both the thermal and mechanical domain.

The test bed provides ideal measurement conditions compared to an assembled machine. It is possible to measure temperature distribution, rotation speed and preload. Additionally the frictional torque of the bearings can be measured directly. This value serves as a reference for the parameter adjustment of the power loss models.

The first tasks of the parameter adjustment process for the model of the bearing test bed are a model analysis (step 1) and a parameter influence analysis (step 2).

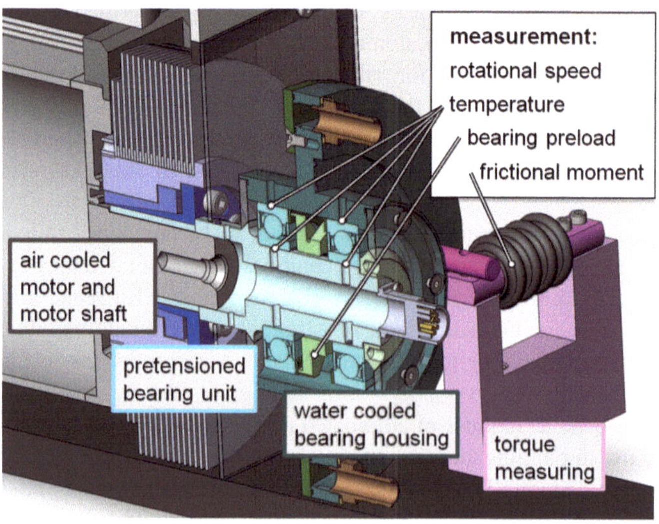

Fig. 13.3 Bearing test bed setup

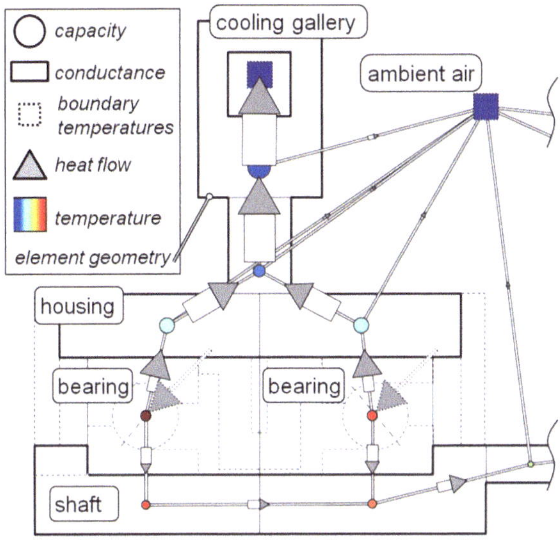

Fig. 13.4 New visualisation method on the example of the analysis of the thermal test bed model and its parameters

Here the effect of uncertain parameters has to be evaluated and adequate measuring variables and positions have to be found. To be able to make statements like these, an overview of the information of the comprehensive and complex model is required. A representation fulfilling these requirements cannot be provided by existing tools of mathematics, digital block simulation or finite element analysis.

To overcome these obstacles, new methods for a condensed visualization of the relevant information have been developed. Figure 13.4 shows the visualisation method for a part of the test bed model. The model structure is represented as a network of the parameters of the thermal model, consisting of capacities and conductances. The Assignment of capacities to the real structure is visualized by the surrounding element contour. Model behaviour is represented by heat flow and temperature and can be animated over simulation time. An innovative aspect of this method is the concurrent representation of qualitative informations by scaling the symbol size of a parameter due to its value. In the example (Fig. 13.4) the parameter values are represented by the symbol area.

The visualisation method allows to recognize qualitative relations between real structure, model structure, model parameters and model behavior at one glance. It enables for important conclusions regarding the thermal behaviour and thus the parameter effect. In this way a rapid and visually assisted formulation of statements relevant for parameter adjustment is possible. The methods were implemented in a software demonstrator.

13.3.2 Load Cases for Data Acquisition

In the sub-task of procedural planning for parameter adjustment (Fig. 13.2, step 3) load cases for the stimulation of the parameter effect on the bearing assembly were developed. The friction model of the bearing involves speed and temperatures as time-variable input parameters. Thus, the uncertain parameters have to be adjusted as a function of these input parameters. Because of nonlinear relations measurement data for the adjustment has to be gathered over the whole range of both input parameters. During the measurement the spindle speed can be adjusted directly using the machine control, whereas the respective temperatures can only be influenced indirectly through friction loss in conjunction with the delayed thermal time response. Thus the needed measurement combinations of temperature and speed have to be achieved through a load case controlled in time.

Several load cases were investigated, thereby intentionally controlling speed values as a function of time according to thermal system properties. The load cases were explored in terms of the coverage of relevant speed and temperature ranges, of the customizing effort of the load templates as well as the time required for the experiments. Figure 13.5 illustrates this on two load cases. Both load cases cover nearly the same speed and temperature range. Load case 1 has a narrow coverage of the parameter range and is easy to customize and execute—but it implies a long experiment time. Load case 2 has a narrow coverage only in selected parameter

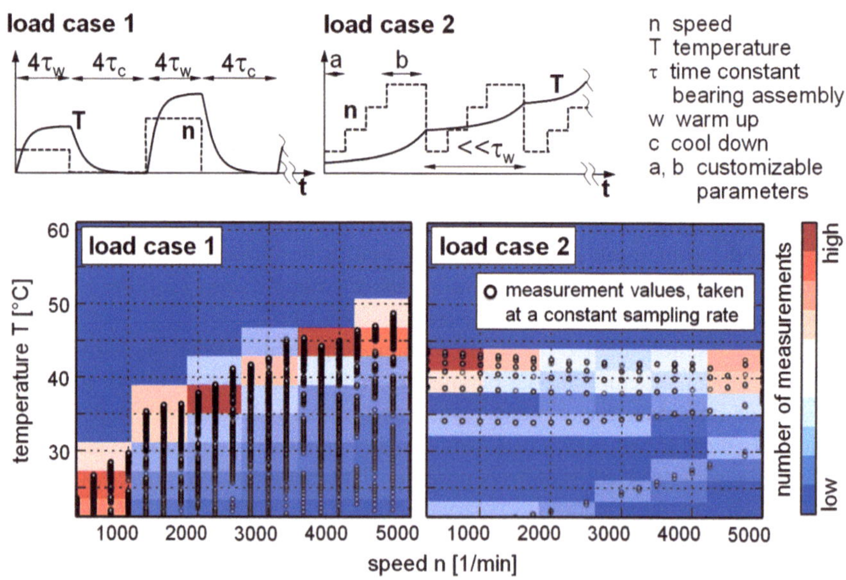

Fig. 13.5 Characteristic of two load templates and the according distribution of measured values (*dots*) with constant sampling rate

ranges and is more difficult to customize and execute. But it implies a far shorter experiment time compared to load case 1 (number of dots).

As a result of these investigations a suitable and time-saving load case (i.e. load case 2) can be developed based on load templates which are tailored to the respective adjustment task. Thus it is possible to estimate the effort and expenditures for measurement and preparation beforehand.

Another topic in the project was the optimization of uncertain parameters on the example of a bearing friction model. The adjustment was based on measured data with a suitable load case chosen. The model according to Palmgren (1964), assuming time-variable parameters of speed n and lubricant viscosity, was used to represent the bearings frictional torque M_R. The parameter of lubricant viscosity is determined by means of a partial model by Vogel-Cameron as a function of temperature T. These partial models contain many constant parameters that have been summed up to the coefficients k_1 to k_4, thus reducing the number of model parameters to be adjusted and leading to the following equation:

$$M_R = k_0 + k_1 \cdot \left(n \cdot e^{\frac{k_2}{T+k_3}}\right)^{k_4} \tag{13.1}$$

The coefficients of the model are the uncertain parameters that were adjusted by means of the Levenberg-Marquardt method as a mathematical algorithm for non-linear optimisation.

Figure 13.6 shows the model with adjusted parameters (grid) and the underlying measured values acquired at the bearing test bed (dots). Initial parameter values were determined by estimates of Schaeffler (2012). The average deviation of the calculated values to the measured values was reduced from nearly 33 % to 10 % after adjustment.

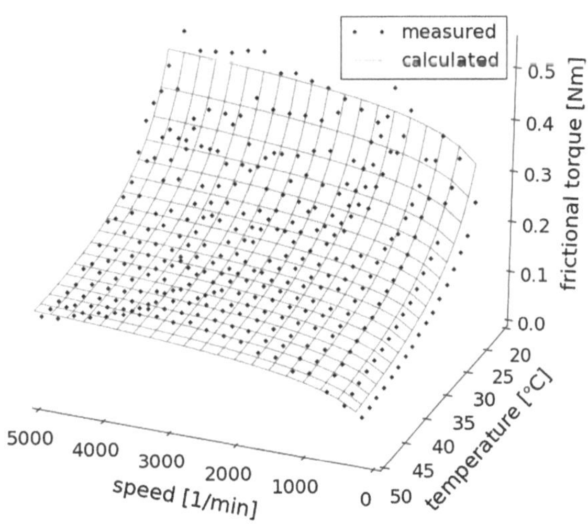

Fig. 13.6 Frictional torque—measured and calculated values based on adjusted parameters

13.4 Summary and Outlook

Thermal models for machine tools include parameters that represent properties varying from machine to machine and over time. These parameters cannot be estimated with sufficient accuracy in the design process. As a result it is necessary to adjust them by measurements. Since the essential properties only emerge during assembly adjustment is only possible in the operational state of the machine tool. On the one hand, great effort and expenditures are required for the process of parameter adjustment. On the other hand, comprehensive specific domain knowledge is needed. As a result, parameter identification can only be conducted under laboratory conditions and is limited to only a few dominant parameters.

Aim of the project is the development of a rapid, comprehensive and efficient adjustment of uncertain parameters. This goal will be achieved by using a systematic approach to support and possibly automate parameter adjustment. In this paradigm activities focus on the development of methods and tools to support the individual steps of parameter adjustment as well as on an integrated data processing concept by linking the tools of the subtasks in a support software.

Methods and tools for higher efficiency in model adjustment were introduced for several sub-steps related to a machine-typical bearing assembly. This development will be continued for further sub-steps. Future activities anticipate refining the algorithms and tools in order to also support more complex application scenarios appearing in axis assemblies and the complete machine.

References

Bäumler St, Brecher C, Haber D, Rossaint J, Großmann K, Kauschinger B, Schroeder St (2012) Simulation der Erwärmung von Spindellagern auf der Basis eines lokalen Reibmodells. 2. Kolloqium zum SFB/TR-96 Thermo-Energetische Gestaltung von Werkzeugmaschinen, Chemnitz, 24–25 Oct 2012

Davidon WC (1991) Variable metric method for minimization. SI-OPT 1(1):1–17

Dementjev A, Ribbecke H-D (2011) Improvement of prediction performance in virtual metrology. In: Proceedings of the 11th European AEC/APC conference. Dresden, pp 71–76

Dibowski H (2013) Semantischer Gerätebeschreibungsansatz für einen automatisierten Entwurf von Raumautomationssystemen. Dissertation, TU Dresden

Dibowski H, Kabitzsch K (2011) Ontology-based device descriptions and device repository for building automation devices. EURASIP J Embed Syst 2011:Article ID 623461

Dibowski H, Kabitzsch K, Ploennigs J (2010) Automated design of build-ing automation systems. IEEE Trans Ind Electron 57:3606–3613

Fritz F (2011) Modellierung von Wälzlagern als generische Maschinenelemente einer Mehrkörpersimulation. Dissertation, Karlsruher Institut für Technologie

Großmann K (2012) Thermo-Energetische Gestaltung von Werkzeugmaschinen – Eine Übersicht zu Zielen und Vorgehen im SFB/Transregio 96. ZWF Zeitschrift für wirtschaftlichen Fabrikbetrieb 107:307–314

Großmann K, Jungnickel G (1998) Das thermische Verhalten veränderlicher Strukturen. Konstruktion 50:27–31

Großmann K, Jungnickel G (2003) Instationäres thermoelastisches Verhalten von Vorschubachsen mit bewegtem Wälzkontakt

Großmann K, Jungnickel G (2006) Prozessgerechte Bewertung des thermischen Verhaltens von Werkzeugmaschinen

Großmann K, Jungnickel G Korrektur thermisch bedingter Fehler an Werkzeugmaschinen mit flexibler Arbeitsraum-Konfiguration. Tagungsband zur CPK 2001

Jedrzejewski J, Kwasny W, Potrykus J (1989) Beurteilung der Berechnungsmethoden für die Bestimmung der Energieverluste in Wälzlagern. Schmierungstechnik 20(8):243–244

Kubin H, Benesch M, Dementjev A, Kabitzsch K, Unkelbach T, Nyderle R, Metzner C (2009) ADM; process identification tool for experts and technologists. In: 35th Annual conference of IEEE industrial electronics, 2009, IECON '09, pp 1444–1449

Marquardt D (1963) An algorithm for least-squares estimation of nonlinear parameters. J Soc Ind Appl Math 11:431–441. doi:10.1137/0111030

Nachimowicz J, Jabrzykowski M (2012) Performance evaluation of rolling element bearings based on tribological behaviour. In: Performance evaluation of bearings. doi:10.5772/51144

Özlük AC (2013) Design space exploration for building automation systems. PhD. dissertation, TU Dresden

Palmgren A (1964) Grundlagen der Wälzlagertechnik, 3rd edn. Franckh'sche Verlagshandlung, Stuttgart

Ryssel U, Distel F, Borchmann D (2011) Fast computation of proper premises. In: Proceedings of the 8th international conference on concept lattices and their applications. Nancy, France, pp 101–113

Schaeffler Technologies (2012) Technische Grundlagen

Chapter 14
Correction Algorithms and High-Dimensional Characteristic Diagrams

Christian Naumann, Ilka Riedel, Ulrich Priber and Roland Herzog

Abstract The time-dependent consideration of the thermal situation in machine tools requires online-enabled methods for the correction of the tool position during the machining process. Higher-dimensional characteristic diagrams, which act as a lookup table for the prediction of the tool center point displacement from several temperature (or other) measurements, are well suited for this purpose. These characteristic diagrams are trained by finite element simulations. On the other hand, least-squares estimators together with reduced order models provide an alternative online-capable prediction technique for the tool displacement. In this context, optimal placement problems for the temperature sensors are described. In any case, the accuracy of simulations and estimation depends crucially on several parameters in the thermo-dynamic model. An approach that reveals the most relevant parameters using sensitivity analysis is also discussed.

14.1 Determination of Relevant Parameters Using Adjoint-Based Sensitivity Analysis

The accurate simulation of the thermo-elastic behaviour of machine tools requires good knowledge of model parameters and boundary conditions, see Table 14.1. Some of these quantities, e.g., material parameters, are often well known, while

C. Naumann (✉) · U. Priber
Fraunhofer IWU Chemnitz, Chemnitz, Germany
e-mail: christian.naumann@iwu.fraunhofer.de

U. Priber
e-mail: ulrich.priber@iwu.fraunhofer.de

I. Riedel · R. Herzog
TU Chemnitz, Chemnitz, Germany
e-mail: ilka.riedel@mathematik.tu-chemnitz.de

R. Herzog
e-mail: roland.herzog@mathematik.tu-chemnitz.de

K. Großmann (ed.), *Thermo-energetic Design of Machine Tools*,
Lecture Notes in Production Engineering, DOI 10.1007/978-3-319-12625-8_14

Table 14.1 Table of symbols

Symbol	Meaning	Unit
T	Temperature	K
u	Displacement	m
σ	Stress	N/m^2
ε	Strain	1
r	Thermal surface load	W/m^2
λ	Thermal conductivity	W/(Km)
ρ	Density	kg/m^3
c_p	Specific heat at constant pressure	J/(kg K)
α	Heat transfer coefficient	$W/(Km^2)$
T_{ref}	Ambient temperature	°C
ν	Poisson's ratio	1
E	Young's modulus of elasticity	N/m^2
β	Thermal volumetric expansion coefficient	1/K

others, such as the heat transfer coefficient describing the thermal interaction between machine and environment, are more difficult to determine. Reliable values for the effective heat transfer coefficient are available in the literature only for simple geometric situations. For complex geometries, one has to resort either to computationally expensive CFD simulations of free convection/radiation problems, or model calibration using experimental data. Both approaches can be rather expensive.

It can be shown that the accurate determination of the effective heat transfer coefficient is not equally important in all parts of the machine structure. Knowledge about less important regions can help to significantly reduce experimental and computational efforts. The importance is based on how strongly inaccuracies in the parameter will cause inaccuracies in the value of the tool center point (TCP) displacement. An adjoint-based sensitivity analysis is proposed to identify regions of particular importance concerning the accurate value of model parameters. This technique can be used for any of the model parameters (or combinations thereof). It is elaborated below in the case of the heat transfer coefficient α.

14.1.1 Background of Adjoint-Based Sensitivity Analysis

The thermo-elastic behaviour of a machine is modeled by a coupled system of partial differential equations. The first part of the system is the heat equation. Here the stationary case is considered,

$$-\operatorname{div}(\lambda \nabla T) = 0 \quad \text{in } \Omega, \tag{14.1a}$$

$$\lambda \frac{\partial T}{\partial n} + \alpha(T - T_{\text{ref}}) = r \quad \text{on } \Gamma. \tag{14.1b}$$

The domain Ω represents the geometry of the machine and Γ is its boundary. The boundary conditions represent a simple model for the effective heat transfer which subsumes the transfer due to radiation and the free convection occurring at the lateral and top surfaces of the machine. The constant T_{ref} denotes the ambient temperature.

The second part of the system is the elastic model, which consists of the balance of forces,

$$-\operatorname{div} \sigma = \mathbf{0} \quad \text{in } \Omega \tag{14.2}$$

involving the stress tensor σ. The relation between stresses and strains is given by Hooke's law. An additive split of the linearized strain tensor ε into its mechanically and thermally induced parts is employed, i.e.,

$$\left.\begin{aligned} &\varepsilon = \varepsilon^{\text{el}} + \varepsilon^{\text{th}}, && \varepsilon^{\text{el}} = \mathfrak{C}^{-1}\boldsymbol{\sigma} := \tfrac{1+\nu}{E}\boldsymbol{\sigma} - \tfrac{V}{E}\,\operatorname{trace}(\boldsymbol{\sigma})I, \\ &\varepsilon(u) = \tfrac{1}{2}(\nabla u + \nabla u^\top), && \varepsilon^{\text{th}} = \beta(T - T_{\text{ref}})I, \end{aligned}\right\} \tag{14.3}$$

which yields an additive split of the stress tensor σ, i.e.,

$$\begin{aligned} \boldsymbol{\sigma} &= \mathfrak{C}\varepsilon - \beta(T - T_{\text{ref}})\,\mathfrak{C}\,I, \\ \mathfrak{C}\,\tau &= \frac{E}{1+\nu}\tau + \frac{E\nu}{(1+\nu)(1-2\nu)}\operatorname{trace}(\tau)I. \end{aligned}$$

The elastic model (14.2)–(14.3) is closed by the boundary conditions

$$u = \mathbf{0} \quad \text{on } \Gamma_D, \tag{14.4a}$$

$$\sigma \cdot n = \mathbf{0} \quad \text{on } \Gamma_N, \tag{14.4b}$$

which represent clamping on a part Γ_D and absence of forces on the remainder of the surface Γ.

This coupled system (14.1a, 14.1b)–(14.4a, 14.4b) can be understood as an implicit equation

$$e(x, \alpha) = 0 \tag{14.5}$$

with states $x = (T, \mathrm{u})$ consisting of the temperature and the displacement, and parameter α.

In order to determine the sensitivity of the state w.r.t. the parameter, the calculation of the derivative of x w.r.t. α is required. This is achieved by total differentiation of (14.5). The directional derivative (sensitivity) of $\alpha \mapsto x(\alpha)$ at a nominal value α_0 of the parameter and under variations in the direction $\delta\alpha$ is denoted by $x'(\alpha_0)\delta\alpha$. This quantity satisfies the following linear equation,

$$\begin{aligned} &e_x(x_0,\alpha_0)x'(\alpha_0)\delta\alpha + e_\alpha(x_0,\alpha_0)\delta\alpha = 0 \\ &\Rightarrow x'(\alpha_0)\delta\alpha = -e_x(x_0,\alpha_0)^{-1}e_\alpha(x_0,\alpha_0)\delta\alpha. \end{aligned}$$

Note that the sensitivity depends on the nominal value (x_0, α_0) for the states and the parameter due to the nonlinear coupling in the boundary condition (14.1b).

As mentioned above, one is not interested in the sensitivity of the entire state but only in the sensitivity of certain quantities such as the TCP displacement. Therefore, one additionally defines an output functional $f : X \to \mathbb{R}$, which maps the state onto the desired quantity. Then the directional sensitivity of the output w.r.t. changes in α follows from the chain rule,

$$\begin{aligned} \frac{\mathrm{d}}{\mathrm{d}\alpha}f(x(\alpha_0))\delta\alpha &= f'(x_0)x'(\alpha_0)\delta\alpha \\ &= -f'(x_0)e_x(x_0,\alpha_0)^{-1}e_\alpha(x_0,\alpha_0)\delta\alpha \end{aligned} \tag{14.6}$$

Evaluating (14.6) *as is* would require potentially repeatedly solving linearized thermo-mechanical models, as represented by the term $e_x(x_0,\alpha_0)^{-1}$ with multiple right hand sides (depending on the dimension of the parameter α).

The main idea of the *adjoint* approach is to evaluate (14.6) in a different order by defining the so-called adjoint state y through the adjoint equation,

$$e_x(x_0,\alpha_0)^\top y = -f'(x_0)^\top.$$

Then (14.6) can be rewritten as

$$\frac{\mathrm{d}}{\mathrm{d}\alpha}f(x(\alpha_0))\delta\alpha = y^\top e_\alpha(x_0,\alpha_0)\delta\alpha. \tag{14.7}$$

It can be concluded that only one (adjoint) thermo-mechanical model has to be solved now. The approach is further illustrated by choosing $f(x) = u_z(TCP)$, i.e., the z-displacement of the TCP. Then the adjoint state $y = (S, v)$, consisting of adjoint temperature S and adjoint displacement v, satisfies the following adjoint system:

$$\begin{aligned} -\mathrm{div}(\lambda\nabla S) &= \beta\frac{E}{1-2\nu}\mathrm{trace}(\varepsilon(v)) \quad &&\text{in } \Omega, \\ -\mathrm{div}\,\tau &= -[0,0,\delta_{\mathrm{TCP}}]^\top &&\text{in } \Omega \end{aligned}$$

with the adjoint stress tensor $\tau = \mathfrak{C}\varepsilon(v)$ and homogeneous boundary conditions

$$\lambda\frac{\partial S}{\partial n} + \alpha S = 0 \text{ on } \Gamma, \quad v = \mathbf{0} \text{ on } \Gamma_D \quad \text{and} \quad \tau \cdot n = \mathbf{0} \text{ on } \Gamma_N.$$

Having solved the adjoint system for (S, v), the desired sensitivity can be efficiently calculated from (14.7), i.e.,

$$\frac{\mathrm{d}}{\mathrm{d}\alpha} f(x(\alpha_0))\delta\alpha = \int_{\Gamma} \delta\alpha(T - T_{\mathrm{ref}})S\,\mathrm{d}s \tag{14.8}$$

for arbitrary, in particular also local, perturbations $\delta\alpha$.

14.1.2 Numerical Results

The method was tested for the Auerbach ACW630 machine column with two heat sources, one at the top of the motor flange and the other at the spindle nut. The picture shows the sensitivity factor $(T - T_{\mathrm{ref}})S$ from (14.8) in (K mm/W). In areas with high values the z-displacement at the TCP is very sensitive w.r.t. changes or inaccuracies in the effective heat transfer coefficient. Hence in these zones the heat transfer coefficient α should be calculated (or determined experimentally) at high accuracy. On the other hand, in areas with low sensitivity values, inaccuracies in α have only little influence on the z-displacement at the TCP.

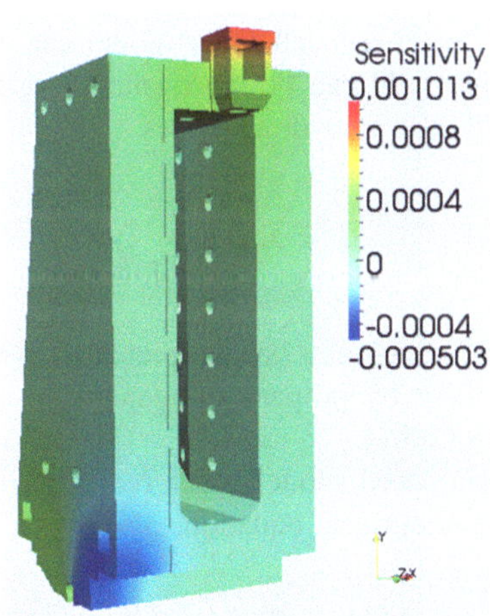

14.2 Optimal Placement of Temperature Sensors for the Estimation of the TCP Displacement

The accuracy and robustness of an estimation of the TCP displacement from temperature measurements depends significantly on the locations where the thermal sensors are placed on the machine surface. This observation is true both for the estimation by means of characteristic diagrams, as in Sect. 14.3, and for a direct estimation method, which is developed in the course of this section.

To be more precise, a simulation-based estimator is developed based on a reduced order model for the TCP displacement, which uses a number of temperature sensors spread across the machine surface. Later on, the sensor positions are optimized with the goal of increasing the robustness of the estimator, i.e., decreasing its sensitivity w.r.t. measurement errors. As in the previous section, numerical results are presented for the Auerbach ACW630 machine column.

14.2.1 TCP Displacement Estimation

With the aim of estimating the TCP displacement, the first goal is to reconstruct the entire temperature field from n_{meas} temperature measurements $\hat{T}_j$, then calculate the displacement field and read out the desired quantity, i.e., the TCP displacement. The reconstruction of the temperature field is in general an ill-posed inverse problem. However, if the operating conditions, especially the heat inputs, of the machine are known, then a reduced order model can be set up a priori. Consequently, the temperature field can be represented as a linear combination of few basis vectors $\varphi_i \in \mathbb{R}^{n_{\text{FE}}}$, also known as modes,

$$T = T_{\text{ref}} + \sum_{i=1}^{n_{\text{POD}}} \zeta_i \varphi_i. \tag{14.9}$$

These basis vectors, which can be interpreted as typical temperature distributions, are determined here by proper orthogonal decomposition (POD), see e.g. Kunisch and Volkwein (2001).

POD is a simulation-based model order reduction method. To apply it, one simulates the time-dependent heat equation,

$$\rho c_p \dot{T} - \operatorname{div}(\lambda T) = 0 \quad \text{in } \Omega, \tag{14.10}$$

with boundary conditions (14.1b) and a possibly time-dependent heat source $r(t)$, and saves snapshots $\mathbb{R}^{n_{\text{FE}}} \ni y_j = T(t_j) - T_{\text{ref}}$ at a number of time points $0 = t_0 < t_1 < \ldots < t_{n_{\text{snap}}}$. The POD approach is to construct orthonormal basis vectors φ_i which approximate these snapshots as well as possible in the least-squares sense, i.e.

$$\text{minimize} \quad \sum_{j=1}^{n_{\text{snap}}} \left\| y_j - \sum_{i=1}^{n_{\text{POD}}} \langle y_j, \varphi_i \rangle W \varphi_i \right\|_W^2, \quad \text{s.t.} \langle \varphi_i, \varphi_j \rangle W = \delta_{ij}, \quad \varphi_1, \ldots, \varphi_{n_{\text{POD}}}.$$

Here W is a finite element mass matrix which represents a suitable inner product. The solution $\varphi_1, \ldots, \varphi_{n_{\text{POD}}}$ of this minimization problem is given by the first n_{POD} eigenvectors of the generalized eigenvalue problem

$$WYY^\top W\varphi_i = \lambda W\varphi_i,$$

where $Y = [y_1, \ldots, y_{n_{\text{snap}}}]$ is the matrix of snapshots.

The reconstruction of the entire temperature field is now reduced to the determination of the few POD coefficients $\zeta_1, \ldots, \zeta_{n_{\text{POD}}}$ such that the reconstructed temperature fits the measurements as well as possible in the least-squares sense,

$$\text{minimize} \quad \frac{1}{2} \sum_{j=1}^{n_{\text{meas}}} \left[T_{\text{ref}} + \sum_{i=1}^{n_{\text{POD}}} \zeta_i \varphi_i(x_j) - \hat{T}_j(x_j) \right]^2, \quad \zeta \in \mathbb{R}^{n_{\text{POD}}}. \tag{14.11}$$

The solution of (14.11) is given by the normal equation

$$J^\top J\zeta = J^\top (\hat{T} - T_{\text{ref}}),$$

where $\hat{T} - T_{\text{ref}} = (\hat{T}_1 - T_{\text{ref}}, \ldots, \hat{T}_{n_{\text{meas}}} - T_{\text{ref}}) \in \mathbb{R}^{n_{\text{meas}}}$ is the shifted vector of measurements and the Jacobian of the residual functions is given by

$$J = J(x) = \begin{bmatrix} \varphi_1(x_1) & \ldots & \varphi_{n_{\text{POD}}}(x_1) \\ \vdots & & \vdots \\ \varphi_1(x_{n_{\text{meas}}}) & \ldots & \varphi_{n_{\text{POD}}}(x_{n_{\text{meas}}}) \end{bmatrix} \in \mathbb{R}^{n_{\text{meas}} \times n_{\text{POD}}}.$$

Using (14.9), one can rewrite the reconstructed temperature as follows,

$$T = T_{\text{ref}} + \Phi(J^\top J)^{-1} J^\top (\hat{T} - T_{\text{ref}})$$

with $\Phi = [\varphi_1, \ldots, \varphi_{n_{\text{POD}}}]$. The TCP displacement can then be calculated by first solving the elasticity Eqs. (14.2)–(14.4a, 14.4b), denoted by the operator $S : T \mapsto \text{u}$,

$$\begin{aligned} u &= ST_{\text{ref}} + S\Phi(J^\top J)^{-1} J^\top (\hat{T} - T_{\text{ref}}) \\ &= \mathbf{0} + S\Phi(J^\top J)^{-1} J^\top (\hat{T} - T_{\text{ref}}), \end{aligned}$$

and then extracting the TCP components from the entire displacement field, i.e.,

$$u(TCP) = \underbrace{BS\Phi\Big(J(x)^\top J(x)\Big)^{-1} J(x)^\top}_{:=A(x)} (\hat{T} - T_{\mathrm{ref}}) \tag{14.12}$$

with an observation matrix B. Note that the matrix $BS\Phi \in \mathbb{R}^{3\times n_{\mathrm{POD}}}$ can be calculated offline by solving only n_{POD} elasticity equations and that during the operation of the estimator, no further solving of the heat equation nor the elasticity system is required. In other words, the reduced order POD model for the temperature naturally induces a reduced order model for the structural mechanics part as well.

The relation (14.12) can be understood as a linear estimator, which maps temperature measurements $\hat{T}$ to the TCP displacement $u(TCP)$. As the Jacobian J depends on the sensor positions $x = [x_1, \ldots, x_{n_{\mathrm{meas}}}]$, the reliability of the estimator depends on x as well.

14.2.2 *Optimization of the Estimator's Quality*

It is well known that the reliability of a least-squares estimator, such as (14.12) can be judged by the eigenvalues of its covariance matrix. Assuming that all temperature sensors have the same standard deviation then the covariance matrix of the estimator is (up to a constant)

$$\begin{aligned} C := \mathrm{Cov}_{\mathrm{TCP}}(x) &= A(x)A(x)^\top \\ &= BS\Phi(J(x)^\top J(x))^{-1}\Phi^\top S^\top B^\top. \end{aligned} \tag{14.13}$$

Large eigenvalues in $\mathrm{Cov}_{\mathrm{TCP}}$ point to a high sensitivity of the estimated TCP displacement w.r.t. perturbations in the temperature measurements, while small eigenvalues stand for a high robustness of the estimator against measurement errors.

This means that a good experimental setup, i.e., a good choice of sensor positions x, is distinguished from a less suitable one by the eigenvalues of $\mathrm{Cov}_{\mathrm{TCP}}(x)$. One possible criterion is the so-called E-criterion, which minimizes the largest eigenvalue of the covariance matrix. To this end, the optimal placement problem could be formulated as follows:

$$\begin{aligned} &\text{Minimize} \quad \lambda_{\max}(\mathrm{Cov}_{\mathrm{TCP}}(x)), \quad x = [x_1, \ldots, x_{n_{\mathrm{meas}}}] \in \mathbb{R}^{3\times n_{\mathrm{meas}}} \\ &\text{subject to} \quad x_1, \ldots, x_{n_{\mathrm{meas}}} \in \Gamma_{\mathrm{feas}}, \end{aligned} \tag{14.14}$$

where Γ_{feas} are the parts of the machine surface where sensors can be placed.

The complexity of the optimization problem (14.14) is based on at least two distinct features. On the one hand, the eigenvalues of $\mathrm{Cov}_{\mathrm{TCP}}$ are a continuous but non-differentiable function w.r.t. the location matrix x, which would require specialized optimization solvers. On the other hand, it would be necessary to parametrize the constraint $x_1, \ldots, x_{n_{\text{meas}}} \in \Gamma_{\text{feas}}$ with a complicated surface such as that of a machine column. In the sequel, the complexity of the problem will be reduced to make it tractable.

First, sensor locations are allowed only in a finite, possibly large, subset Γ_{finite} of Γ_{feas}. An obvious choice for Γ_{finite} would be the nodes of the finite element mesh. The placement problem then becomes a high dimensional combinatorial optimization problem. For the machine column it is $|\Gamma_{\text{finite}}| = 25{,}288$ and there would be 319,728,828 possibilities to place only two sensors. To this end, the simultaneous placement of all sensors as in (14.14) is replaced by a sequential (greedy) placement of one sensor at a time, see Algorithm 1. This will lead to sub-optimal solutions of (14.16), but render the overall problem tractable. Numerical evidence in Herzog and Riedel (2013) justifies this approach.

Algorithm 1 (Sequential placement)

1: *Set* $i := 0$
2: ***repeat***
3: *Set* $i := i + 1$
4: *Solve problem* (15)
5: ***until*** $i \geq n_{POD}$ *and stopping criterion* (17) *is met*

In each step of the sequential placement the position of one sensor is determined by solving the optimization problem (14.15), while other sensors at positions $x_1, \ldots, x_{i-1}$ have already been placed and will be kept fixed.

$$\begin{aligned} &\text{Minimize} \quad \lambda_{\max}(\mathrm{Cov}_{\mathrm{TCP}}(x_i)) \\ &\text{subject to} \quad x_1 \in \Gamma_{\text{feas}}. \end{aligned} \tag{14.15}$$

Herein,

$$\mathrm{Cov}_{\mathrm{TCP}}(x_i) = BS\Phi_i(J_i(x_i)^\top J_i(x_i))^{-1}\Phi_i^\top S^\top B^\top \tag{14.16}$$

denotes the covariance matrix of the estimator in stage i, which maps the first i temperature measurements to the TCP displacement vector. Herein, J_i and Φ_i are restrictions of J and Φ, see Herzog and Riedel (2013) for details.

Besides making the placement tractable, the sequential placement approach allows the determination of the number of sensors n_{meas} on the fly rather than a priori. Algorithm 1 is stopped as soon as $i \geq n_{\text{POD}}$ and

$$\lambda_{\max}(\mathrm{Cov}_{\mathrm{TCP}}(x_i)) \leq \left(\frac{\varepsilon_{\text{tol}}}{\sigma r(\alpha)}\right)^2 \tag{14.17}$$

with a user-defined tolerance ε_{tol}, the standard deviation σ of the sensors, a confidence level $\alpha \in [0, 1]$ and the squared radius $r(\alpha)^2 = F_{\chi_3^2}^{-1}(\alpha)$ the inverse cumulative distribution function of a χ^2 distribution with three degrees of freedom. The condition

(14.17) guarantees that the largest half axis of the confidence ellipsoid describing the estimated tool center point displacement will be at most ε_{tol}.

14.2.3 Numerical Results

The sensor placement was tested again for the Auerbach ACW630 machine column. Two heat sources were assumed, one on top of the motor flange and the other at the spindle nut, see Fig. 14.1a. The TCP displacement was approximated by the mean displacement of the mounting points shown in Fig. 14.1b of the sledge holding the spindle with the tool, shown in Fig. 14.1b.

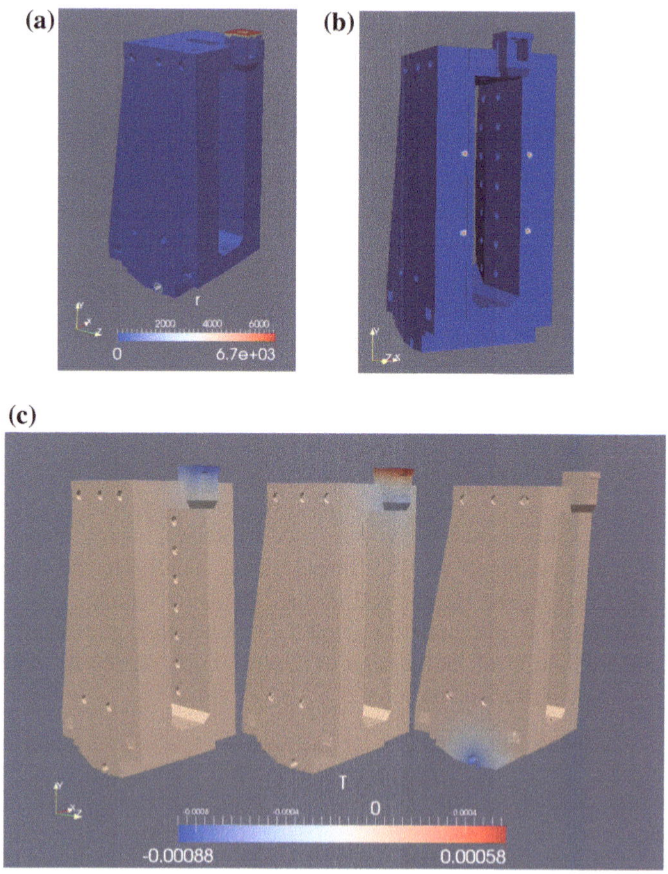

Fig. 14.1 Geometric location of heat sources, mounting points and first POD modes. **a** Location and strengths in W/m^2 of the heat sources r at Γ_{flange} and Γ_{nut}. **b** Mounting points which determine the TCP location. **c** First three POD modes φ_i $i = 1, 2, 3$

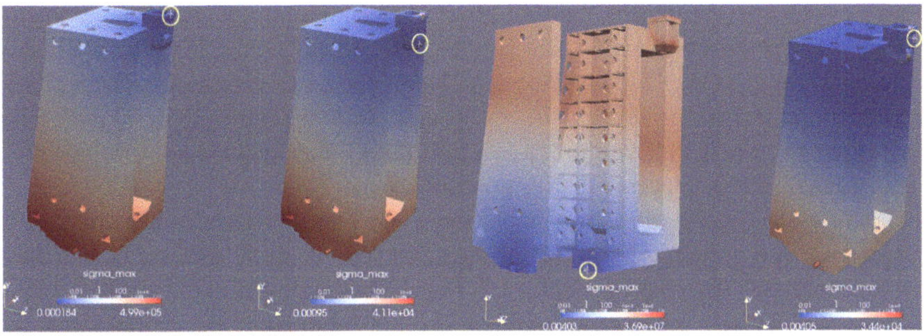

Fig. 14.2 First four sensor positions

A POD simulation was carried out by heating each of the two sources until a stationary state was reached and cooling it down to the ambient temperature afterwards. It turned out that $n_{\mathrm{POD}} = 5$ basis vectors, Fig. 14.1c shows the first three modes, yield a pretty good reduced model for the temperature.

For the constants in the stopping criterion (14.17) of algorithm 1 a standard deviation of $3\sigma = 0.1\mathrm{K}$ was chosen, i.e., 99.73 % of all temperature measurements lie within ± 0.1 K of their true value, a confidence level of $\alpha = 0.95$ and a tolerance $\varepsilon_{\mathrm{tol}} = 1 \times 10^{-6}$ m. Then one obtains that $n_{\mathrm{meas}} = 6$ sensors are enough to reach the desired accuracy of $\varepsilon_{\mathrm{tol}}$. Figure 14.2 shows the square root of the maximal eigenvalue of the covariance matrix (14.16) during the first four placement problems, as well as the optimal sensor locations (yellow circles) at the points where the largest eigenvalue is minimal. As expected, all sensors lie near the two heat sources at the motor flange and the spindle nut.

Placing sensors in this way, one obtains a Jacobian $J(x) \in \mathbb{R}^{n_{\mathrm{meas}} \times n_{\mathrm{POD}}} = \mathbb{R}^{6 \times 5}$, which leads (in a suboptimal sense due to the sequential placement) to a TCP displacement estimator (14.12) with minimal variance in the sense of (14.14).

14.3 Characteristic Diagrams

An alternative way of estimating the TCP displacement from temperature measurements according to (14.12) is through the use of characteristic diagrams. This can be conceived as a multi-dimensional lookup table which has been set up and trained in an offline phase. Typical input dimensions include temperature measurements at certain locations, but might include other information, such as the current geometrical configuration, (e.g. the current spindle height), as well. In this respect, characteristic diagrams are more flexible than direct estimators such as (14.12). The output in the present context would consist of the three components (x, y, z) of the TCP displacement.

Effectively, the use of a characteristic diagram corresponds to an interpolation and extrapolation of thermal states from selected locations (or selected grid points in a FE mesh) established during the training phase. Depending on the machine geometry and the FE discretization used, there will be many geometrical grid points with very little significance to the TCP displacement, either because their temperature is highly dependent on that of their neighbors, or because deformations in their vicinity have little effect on the TCP. Therefore, a few grid points are expected to be sufficient to estimate the TCP displacement with good accuracy.

Finding out which grid points matter most and how many are at least needed to achieve a given accuracy is subject of the sensitivity analysis and can be achieved with the methods described in Sect. 14.2. For now, it is simply assumed that a small number n_g of such grid points suffices for the prediction. By determining the maximum and minimum temperature each grid point may reach during the machine tool's operation and splitting the resulting intervals into discrete segments one gets n_g 1D grids, which are combined into one large n_g-dimensional grid of temperature variables. Each point in this grid represents one thermal state to which a TCP displacement vector is assigned.

$$f : \mathbf{t} = \left(T_1, T_2, \ldots, T_{n_g}\right) \in \mathbb{R}^{n_g} \rightarrow (x, y, z)_{\mathrm{TCP}} \in \mathbb{R}^3$$

Using data gathered in experiments or obtained through simulation, an attempt was made to fit an n_g-dimensional hypersurface into this $(n_g + 1)$-dimensional space, see Fig. 14.3. As with any fitting, these samples need to cover as much of the grid as possible. Especially with measured data points one also needs some redundancy to reduce the effect of random errors.

A method called Smoothed Grid Regression (SGR) that was developed prior to this project, see Priber (2003) and Naumann (2012), uses least squares to calculate such hypersurfaces for a given temperature grid and a database of samples. It does so by weighting the penalties resulting from data errors, i.e., the distance of the samples from the hypersurface, against those resulting from a lack of smoothness, i. e., from large changes in the deformation gradient. This will lead to a large and over-determined linear system such as

$$\begin{pmatrix} \mathbf{z} \\ \mathbf{0} \end{pmatrix} = \begin{pmatrix} A \\ S \end{pmatrix} R,$$

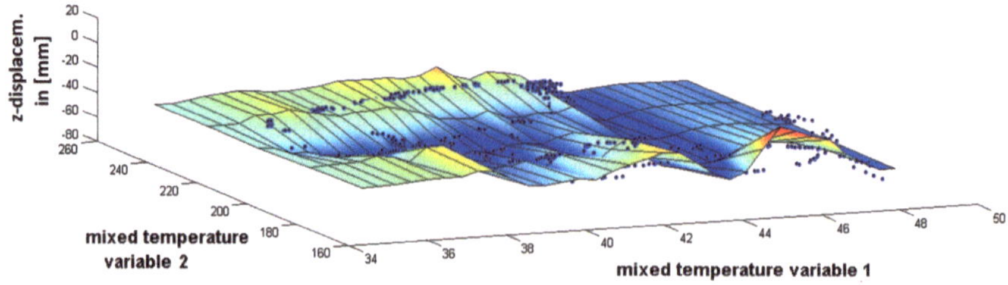

Fig. 14.3 Characteristic diagram for two temperature inputs

where each column of A and S corresponds to a grid vertex. Each row of matrix A describes the location of a data point w.r.t. its surrounding grid points while its value is the corresponding entry in $\mathbf{z}$. Each row of matrix S requires a grid point's value to be colinear to its left and right neighboring values in every grid dimension.

While the SGR method requires very little knowledge of the underlying problem, its approximation quality heavily depends on the choice of input variables used for the grid. Another weakness of this method is the exponential growth of the resulting linear system with the dimension of the grid. Even though the assembled matrices have a very sparse structure, with only few elements in each row and column, problems with memory shortage arise quickly as the grid dimension increases. If direct solvers are used to solve these linear systems (in a least-squares sense) then the unavoidable fill-in will cause an overflow much sooner still. The same goes for the computation time, which grows along with the grid size. The main goal is therefore to make very large least-squares problems computable in order to boost the approximation quality of characteristic diagrams.

To solve these performance issues three elements are needed: a fast iterative solver, a way of solving the linear system without having to assemble the entire matrix and a way of reducing a large complete grid to a much smaller grid without much loss of accuracy. While there are a number of fast direct solvers, some of which make very good use of the SGR's band matrix structure, they are bound to create a certain fill-in which grows rapidly for an increasing grid fineness. They also have trouble using the underlying grid and with assembling the matrix piece-by-piece during the solution process. For all these reasons iterative solvers become attractive alternatives. Since the efficiency of iterative solvers for symmetric problems (such as least-squares) largely depends on the condition number and distribution of eigenvalues, preconditioning becomes an important aspect. The SGR is a grid based method where the fineness of the grid can be arbitrary. Therefore, it is easy to establish a set of hierarchical grids that can be used by a multigrid solver or a multigrid preconditioner, e.g., in a conjugate gradient method.

Multigrid methods, see e.g. Trottenberg et al. (2001), were originally developed to solve linear systems arising through the discretization of a partial differential equation (PDE), e.g., from a finite element approach. By expressing the regression problem as an energy minimization problem

$$\min_u \quad \frac{1}{2}\sum_{i=1}^{n_{\text{data}}} |z_i - u(\mathbf{x}_i)|^2 + \frac{1}{2}\int_\Omega \lambda(\mathbf{x}) \cdot |\nabla u(\mathbf{x})|^2 \mathrm{d}\mathbf{x},$$

one gets the following PDE in weak form:

$$\forall \delta u: \quad \sum_{i=1}^{n_{\text{data}}} (z_i - u(\mathbf{x}_i)) \cdot \delta u(\mathbf{x}_i) + \int_\Omega \lambda(\mathbf{x}) \cdot \nabla u(\mathbf{x}) \cdot \nabla \delta u(\mathbf{x}) \mathrm{d}\mathbf{x} = 0.$$

The second term in the energy functional corresponds to a smoothing term. The one shown above prefers constant values of the unknown function u, whereas terms favoring constant gradients are also used frequently.

Experimentally it was shown that this SGR-based multigrid preconditioned CG solver achieves optimal complexity. Even the best iterative solver cannot beat $\mathcal{O}(n_p)$ operations, where n_p is the total number of vertices in the temperature grid. It is thus essential to make the grid as small as possible for a given, fixed accuracy. Using an orthogonal grid is quite useful both in the multigrid and in the FEM but a simple grid in which every grid axis corresponds to one geometrical node leads to a hypercube of thermal states in which large parts can never be realized during the machine tool's operation. This problem calls for a principal component analysis of the input variables (temperatures). By rotating the hypercube and shrinking it to form a kind of minimal orthogonal convex hull around the relevant thermal states, one accomplishes the task of concentrating most of the information in only a few wide dimensions, which makes it possible to cut off all the very narrow dimensions and to use a coarser discretization on the intermediate ones.

Another technique for reducing the number of grid points involves the use of adaptive FEM. Rather than discretizing the entire hypercube uniformly down to the finest level, one can also make the discretization finer in those regions with the greatest data error. In this particular case these are the states with large second derivatives of the TCP displacement $u(\mathbf{x})$.

In the following test performed on experimentally measured data, the results of the described characteristic diagram approach are demonstrated. To do this a training data set and a test data set were assembled from two measurements on a machine tool. Figure 14.4 shows how well the characteristic diagram based on the training data predicts the displacements of the test data.

One can see that the predicted TCP displacement (thin, black) roughly follows the actual displacement (thick, gray). While the training data (not shown here) can be predicted with no more than ±10 μm, the test data still produces a maximum

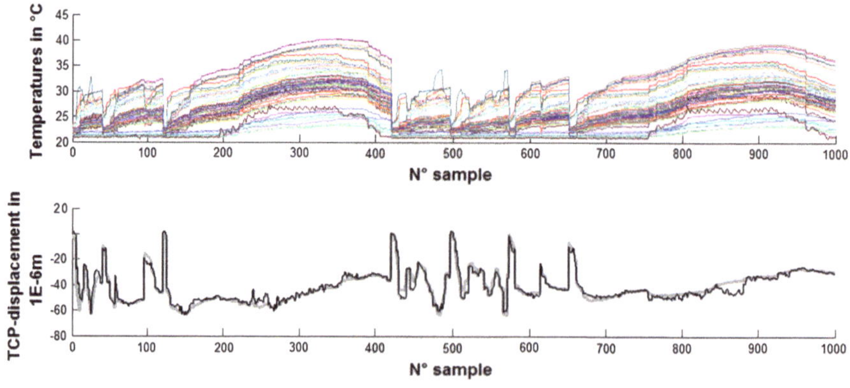

Fig. 14.4 Approximation of TCP displacement from temperature sensor data

error of ±20 μm. This reduces the maximum error to one third, which is a good start. Parts of the predicted curve that strongly deviate from the measured curve suggest that the test data contradicts the training data. This most likely means that some data points that lie in the same region of the grid strongly differ in a dimension that cannot be seen. One can therefore not fully distinguish all relevant thermal states of the machine tool from this thermal sensor setup, which shows the importance of the choice of input variables for the quality of any characteristic diagram. Similar tests have shown that a poor choice of input variables leads to particularly large errors when the temperature gradient is big. In these cases some machine tool parts have already heated up, expanded and thus caused a TCP displacement before any of the thermal sensors have picked it up. The opposite case has also occured, where the thermal sensors have picked up heat before any of the relevant machine parts have started expanding.

Beside further advances in the efficient calculation of characteristic diagrams, another interesting research topic will be the improvement of predictions with a non-ideal choice of input variables. This will be particularly relevant in cases in which the sensor placement is restricted by safety or practical limitations.

14.4 Integration into the CRC/TR 96 and Outlook

To summarize, subproject B05 is one out of three subprojects dealing with real-time correction algorithms, where the time scale is determined by the thermal constants of the machine tool. All three approaches are based on a prediction of the thermally induced TCP displacement in order to allow for its correction through the machine's NC controller.

The three approaches differ with respect to the input data that serve as the basis of the TCP displacement prediction. They also differ with respect to the way in which the thermo-mechanical functional chain is taken into account. Subproject B05 offers two algorithms to achieve the prediction of the TCP displacement. First, characteristic diagrams (Sect. 14.3) capture the thermo-mechanical behavior through a multidimensional lookup table which needs to be set up in an offline phase either from measurements or simulated data. Second, a least-squares estimator (Sect. 14.2) has been developed that incorporates a reduced order model of the thermo-mechanical effects. In contrast to B06 and B07, both algorithms typically use temperature measurements, which are easy to gather, as their input, but are open to extension to other input data. Since the temperature data constitute part of the state of the machine (in contrast to, e.g., thermal loads), both algorithms in B05 avoid a drift-off of the predicted versus the true thermal state due to model inaccuracies. A final distinguishing feature of the approach taken in project B05 is the optimal placement of temperature sensors according to their information content with regard to the TCP displacement. This helps to reduce the amount of input data and further adds to the robustness of the estimator.

It should be mentioned that all results so far were obtained for a static pose of the machine tool and it is ongoing work to incorporate changes in the pose. Moreover, we emphasize that all projects that employ simulation technology, among them A05, A06, A07, B05, B07, rely on the accuracy of model parameters for realistic results. Most notably, this concerns the heat transfer coefficient, which determines the rate of heat exchange with the machine's environment. As shown in Sect. 14.1, however, the TCP displacement may depend strongly on the actual value(s) of this parameter in certain regions of the surface. Therefore, the real-time identification of the heat transfer coefficient will be a major goal for the future work in this project.

References

Herzog R, Riedel I (2013) Sequentially optimal sensor placement in thermoelastic models. Technical report, TU Chemnitz

Kunisch K, Volkwein S (2001) Galerkin proper orthogonal decomposition methods for parabolic problems. Numer Math 90(1):117–148

Naumann C (2012) Proceedings, 22nd workshop computational intelligence, Dortmund, 6–7 Dec 2012

Priber U (2003) Proceedings, 13th workshop fuzzy system 2003, Dortmund, 19–21 Nov 2003

Trottenberg U, Oosterlee CW, Schüller A (2001) Multigrid. Academic Press Inc., San Diego, CA, 2001 (With contributions by A. Brandt, P. Oswald and K. Stüben)

Chapter 15
Correction Model of Load-Dependent Structural Deformations Based on Transfer Functions

Christian Brecher, Marcel Fey and Matthias Wennemer

Abstract An indirect technique for the correction of thermo-elastic displacements induced by loads of the linear axes is developed in subproject B06. The basis is formed by an control-internal correction approach describing the functional relationship between the mechanical load profile and the displacement of the Tool Center Point using transfer functions of different orders. The model parameters referred to are determined in displacement experiments. Since axes load-induced displacements highly depend on the position, a volumetric measurement and correction approach is utilised.

15.1 Introduction

The principal aim of the subproject B06 is a method for correcting thermally induced machining errors based on control-internal data, independent of the type and amount of machine load and the Tool Center Point (TCP) position. This control-internal correction results in both high precision machining and energy efficiency, since cooling and tempering measures can be reduced and remachining becomes unnecessary.

The correction method involves load-induced thermo-elastic displacements of the TCP in the entire workspace, resulting from the load acting on the linear axes, as well as the spindle. The correction of spindle load- and ambient temperature-dependent thermally induced TCP displacements, that have already been developed at the

C. Brecher · M. Fey · M. Wennemer (✉)
Laboratory for Machine Tools and Production Engineering (WZL), RWTH Aachen University, Aachen, Germany
e-mail: m.wennemer@wzl.rwth-aachen.de

C. Brecher
e-mail: c.brecher@wzl.rwth-aachen.de

M. Fey
e-mail: m.fey@wzl.rwth-aachen.de

K. Großmann (ed.), *Thermo-energetic Design of Machine Tools*, Lecture Notes in Production Engineering, DOI 10.1007/978-3-319-12625-8_15

Laboratory of Machine Tools and Production Engineering (WZL) of the RWTH Aachen, are associated with the newly developing axes dependent model.

Measurement of workspace-dependent displacements is conducted by means of tracking interferometers, which are commonly used in coordinate measuring technology, but nowadays have also been introduced in machine tools for volumetric corrections. This measuring equipment makes it technical feasible to quickly record the workspace dependent displacements for a great variety of load states in the micrometer range. Since modeling individual positions in the workspace would be inefficient for a thermal correction, a method according to the error parameters of the ISO 230 has been designed. This makes it possible to derive the thermal displacements over the entire workspace based on just a minimum of measuring points.

15.2 Approach

The development focuses on a transfer function based correction procedure for axes, load-dependent thermo-elastic displacements. In contrast to spindle related thermo-elastic displacements, those of the axes depend to a great degree on the position. Consequently, the entire workspace has to be considered.

15.2.1 Correction Method

The correction algorithm bases on transfer functions between the load-representative physical auxiliary variables regarded as the cause and the TCP displacement regarded as the effect. In the easiest case, modelling is carried out using delay elements of the PT1 form. The associated functional relationship is characterized by the following differential equation in the time domain:

$$T \cdot \dot{y}(t) + y(t) = K \cdot u(t) \quad \text{whereby} \tag{15.1}$$

$y(t)$ output variable (displacement)
$u(t)$ input variable (power)
K amplification factor
T time constant

Consequently, the complex transfer function in the Laplace domain follows the expression:

$$G(s) = \frac{K}{1 + T \cdot s} \tag{15.2}$$

The order and quantity of transfer functions for the correction algorithm have to be chosen appropriately on basis of the experimental determined thermo-elastic displacement characteristics of the machine tool. Afterwards, the correction algorithm is formulated for the selected transfer functions and the determined parameters are stored in the machine tool control. The limitation of the experimental scope to a few representative load cases that are capable of representing the thermal axis characteristics is decisive for implementing the correction technique on an industrial scale.

Figure 15.1 elucidates the integration of the correction method into the machine tool. In future operation, the correction technique running at the Human Machine Interface (HMI) records the required machine parameters, such as power or revolutions per minute, and calculates the TCP's displacement for fixed or variable times. The minute range has to be defined by means of the stored correction algorithms. At the end, the calculated displacement values are passed to the machine control as an offset.

As a rule, each variable available by machine control with a functional relationship to the displacement can be considered as an auxiliary variable. The calculated power of the associated axis, which is already been provided by the Numerical Control (NC), is a suitable variable. However, for axes related errors at least one second parameter is required to distinguish between a low traverse speed with a high cutting force and a high traverse speed with a low cut-ting force. Thus, it can be expected that, in the latter case, the guideways will heat up much more and, consequently, the thermo-elastic deformations of the machine structure will differ. Speed, which—like power—can be directly read out from the machine control, is a parameter suitable for this distinction.

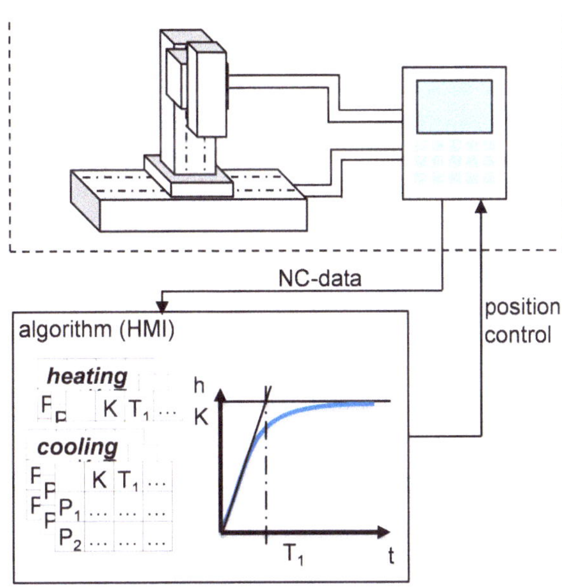

Fig. 15.1 Correction approach

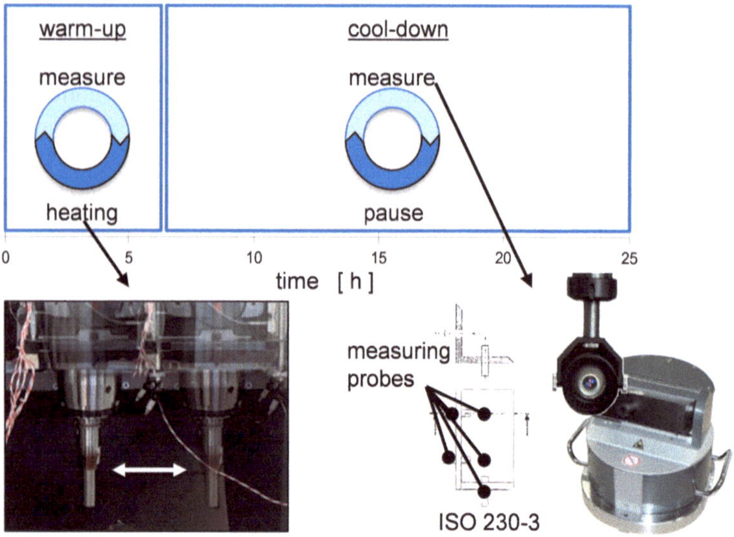

Fig. 15.2 Experimental design

15.2.2 Experimental Methodology

Knowledge of the step function in form of the load and the step function response as the TCP displacement is sufficient to suitably parameterize the transfer functions. For this reason, displacement experiments of the machine tool axes are conducted under constant load to model the thermal characteristics.

Figure 15.2 illustrates the sequence of experiments to investigate the thermo-elastic characteristics. The experiments consist of both a heat up and a cool down stage by reason of different time constants which requires individual modeling. During experiments, the TCP displacements are continuously captured in time intervals to be defined as a function of the machine behavior, together with NC data, such as power and revolutions per minute. Displacements have been previously measured point-wise using a fixture with displacement transducers according to ISO 230-3 for the preliminary investigations, as well as in a volumetric manner by means of tracking interferometers. Based on these measurements, it is possible to determine the parameters of the transfer functions.

15.3 Results

The idea for a transfer function based correction model has already been realized and successfully experimented for the spindle load-induced displacement error at the WZL as part of the EU project entitled “Next Generation Production Systems”.

In this process, the thermal spindle displacements are modeled for one point in the workspace under various loads by a specifically designed loading unit. It could be demonstrated that thermo-elastic TCP displacements can be reduced by 70 % using the existing correction approach (Brecher and Wissmann 2009).

Subproject B06 is aimed at transferring this methodology to the axes portion of thermally induced displacements and thus due to position dependency of axes related errors into the entire workspace.

15.3.1 Stressing Unit for a Targeted Load of the Machine Axes

Axis-related heat transfer into the machine structure caused by a load resulting in a displacement of the TCP position can mainly be traced back to tribologic effects (e. g. rolling bearings, guideways etc.) and electrical energy losses that appear during cutting and air cuts to differing extents. Therefore a stressing unit was designed to investigate the influence of these effects by experiments. The unit is capable of charging the linear machine axes separately and simultaneously with a defined and reproducible load. The energy loss that emerges within the structure of the stressing unit has to be isolated from the machine to prevent foreign influences. In addition to these requirements, measurability of the entire machine volume must be particularly maintained in order to apply volumetric measurements with sufficient accuracy.

Figure 15.3 illustrates the engineered stressing unit with the corresponding hydraulic circuit diagram. The entire unit consists of two separate linear axes with a ball screw. They are attached to a cross table by means of a bracket and mounted as a parallel kinematic to the machine tool. To avoid distortions between the kinematic mechanisms, the complete unit is linked via a ball joint at the spindle. Furthermore an automatic uncoupling and recoupling of the unit for the measuring procedure is possible.

In this setup, the load is charged using a hydraulic pump, working against a pressure control valve. It is mechanically driven by means of the ball screw of the stressing unit via the machine tool. The easy dissipation of heat losses is in comparison to others an advantage of hydraulic systems. To charge a defined load, a force measurement device for all three spatial directions is positioned between the spindle and the stressing unit.

15.3.2 Displacement Model for One Point in the Workspace

Prior to volumetric considerations, a displacement model describing one point in the workspace is created.

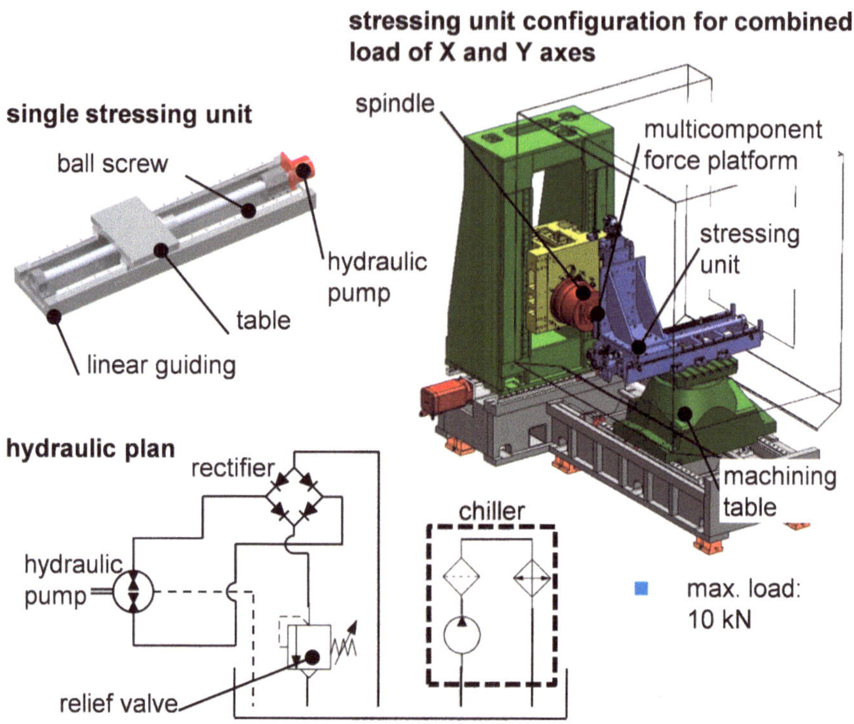

Fig. 15.3 Stressing unit

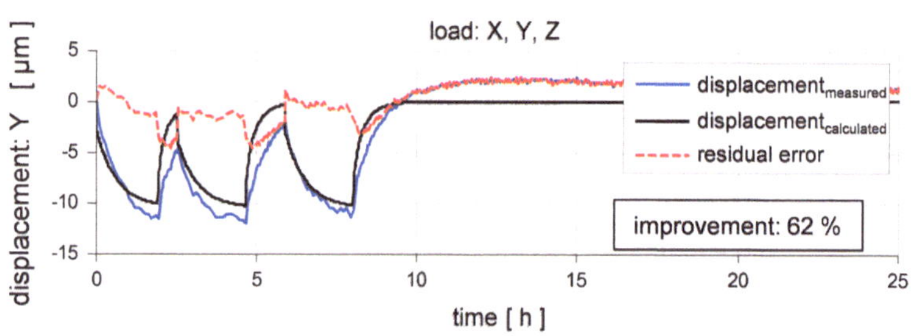

Fig. 15.4 Correction results for a specific point of the machine volume (Brecher and Wennemer 2013)

The result of an validation cycle of this model is shown in Fig. 15.4. In this experiment, displacements are measured according to ISO 230-3 by means of a fixture and five inductive displacement transducers (see Fig. 15.2).

The model used for this purpose based on two experiments per axis, which were conducted using two different speed/power values over 24 h per experiment. In each of these modelling trials only one axis is loaded. The TCP position's displacement of the single axis is represented by an PT1 approach. As a result, six displacement experiments provide the database for modelling the investigated 3 axis machine. During validation trial of Fig. 15.4 all axes are loaded simultaneously. Furthermore both, load and off periods and motion sequences varied.

As a result, it is possible to reduce the displacement in the validation trials to less than 5 μm when using the actual conditioned model (Brecher and Wennemer 2013). Reliable results are obtained during heat up, but further optimization efforts are needed for the cool down phase. The chosen method of a transfer functions based correction of axial errors is thus promising.

15.3.3 Development of a Volumetric Method to Measure Thermo-Elastic Displacements

As has been mentioned, thermo-elastic displacements regarding the axes are position-dependent, so that modelling has to involve the entire machining space. Thus, a volumetric method of measurement has to be drafted and experimented for thermo-elastic displacements. Furthermore the existing modelling method has to be applied to the entire machine volume.

Tracking interferometers by the firm Etalon (Schwenke et al. 2008) are used to measure the thermo-elastic displacements over the entire machine tool volume. In contrast to laser trackers, the measuring method using the so-called Laser Tracers is distinguished by the fact that only the length data, and no angular data at all, is applied to determine the relative reflector positions. Thanks to this analysis method and the internal structure the influence of angular errors of the Laser Tracer motor position on the measurement result is negligibly low. Figure 15.5 shows the measurement system in use.

With the conventional measurement and analysis method, the relative mutual location of the spindle positions can be determined in the micrometer range. For a thermally induced displacement, the position of the entire volume relative to the machine table—in contrast to a solely geometric correction—is also relevant.

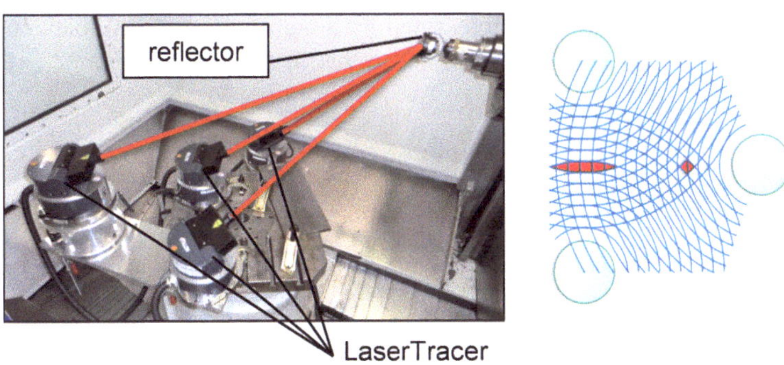

Fig. 15.5 Volumetric measuring method

However, this position is characterised by substantially higher measurement inaccuracies. These uncertainties result from the high inaccuracy in determining the Laser Tracer positions, whose relative location is the referencing connector between the single measurements. They are a function of the measurement strategy and volume and approximately reach 15 μm. However, the measurement inaccuracy specified can be reduced by about 50 % including a time series analysis involving all series of measurement. Due to the fact of a referencing connector the relative position of the Laser Tracers to each other has to be guaranteed by a thermostable linkage mechanism made of, for instance, carbon fibre-reinforced plastics. Furthermore appropriate software was developed for automatic measurement and the necessary calculations. Thus, a measurement method with satisfactory precision for the correction of thermo-elastic displacements is made available (Brecher et al. 2014).

15.4 Classification of Outcomes CRC/TR 96

The introduced correction method offers the potential to correct axes related thermally induced machining errors over the entire workspace of machine tools. This method could enable highly accurate machining to be efficient in the sense of the global objective of the CRC/TR 96, since re-work is unnecessary.

The subproject B07 maps the whole thermo-energetic functional chain for the generation of thermo-elastic errors in a structure model. This means that the final displacement is derived from the mechanical load profiles as input variable by very detailed thermal models. The difference is that, in the transfer function based correction, this functional relationship between the displacement and the load is represented in a simplified manner by a model based on delay elements. Modelling effort for a detailed virtual image becomes unnecessary, on the one hand. On the other hand, the experimental efforts to identify the parameters expand. In contrast to these two techniques, subproject B05, characteristic diagram based correction, uses measured temperature fields in the machine structure together with their corresponding TCP-displacements as input variables. The displacement's functional dependency on temperature is represented by a higher-dimensional map, and thus entails a high level of abstraction similar to that in subproject B06.

15.5 Outlook

The first trials to validate the transfer function based correction method for axes related thermally induced displacements in a workspace point lead to promising results. Furthermore, it is now possible to measure the displacements in the volume in an automated manner by means of Laser Tracers. Now the modelling method can be transferred to the entire operating volume as part of the current activities.

References

Brecher C, Fey M, Wennemer M (2014) Volumetric compensation of thermal machine tool errors as a result of linear axes loads. Special Interest Group: Thermal Issues

Brecher C, Wennemer M (2013) Eigenschaftsmodellbasierter Ansatz zur Korrektur thermoelastischer Verlagerungen. Tradition und Gegenwart bei der Analyse des thermischen Verhaltens spanender Werkzeugmaschinen. Dresden, pp 155–170

Brecher C, Wissmann A (2009) Modelling of thermal behaviour of a milling machine due to spindle load. In: 12th CIRP conference on modelling of machining operations, vol 2, pp 673–678

Schwenke H, Knapp W, Haitjema H et al (2008) Geometric error measurement and compensation of machines—an update. CIRP Ann Manufact Technol 57:660–675. doi:10.1016/j.cirp.2008.09.008

Chapter 16
Structural Model-Based Correction of Thermo-elastic Machine Tool Errors

Xaver Thiem, Knut Großmann and Andreas Mühl

Abstract With the load data collected in a machine tool control, the resultant thermally induced error at the tool center point (TCP) can be concluded by means of the thermal and thermo-elastic simulation models derived from the machine tool structure, while the machining accuracy can be increased upon correction of this error. This paper provides an overview of the software modules required to implement the structural model-based correction. The main modules are detailed in terms of their characteristics (such as interfaces and cycle times). The concrete implementation of the communication interface between the control core and the model level is briefly demonstrated for the Beckhoff TwinCAT3 control and the simulation software Matlab.

16.1 Introduction

In subproject B07, a method for the structural model-based correction of thermo-elastically induced machine tool errors is generated. In this correction method, the load and trajectory data logged in the control (motor current and velocity values, positions) are converted to heat fluxes in motors, bearings and guidances by means of power dissipation models. The load data are converted into thermal conduction values, which map the heat transfer between these moving assemblies. These heat fluxes and heat transfer representations are specified in a thermo-elastic network model. The network model consists of reduced-order FE models describing structural

X. Thiem (✉) · K. Großmann · A. Mühl
Faculty of Mechanical Engineering, Institute for Machine Tools and Control Engineering, Technical University Dresden, Dresden, Germany
e-mail: xaver.thiem1@tu-dresden.de

K. Großmann
e-mail: knut.grossmann@tu-dresden.de

A. Mühl
e-mail: andreas.muehl@tu-dresden.de

K. Großmann (ed.), *Thermo-energetic Design of Machine Tools*,
Lecture Notes in Production Engineering, DOI 10.1007/978-3-319-12625-8_16

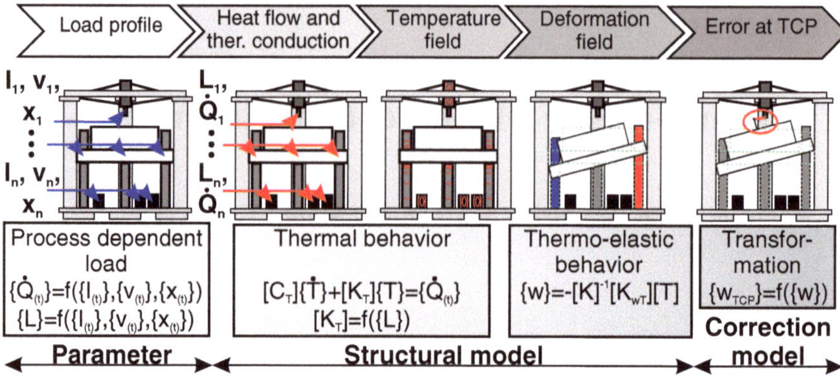

Fig. 16.1 Thermo-elastic functional chain shown for the example of the “MAX” test bed

machine parts. They are connected with thermal conduction values, which represent the guidance, contact and surrounding conditions. The network model is numerically integrated in the time domain. The error at the tool center point (TCP) is calculated as a function of position through this numerical integration and follow-up deformation calculation and is then sent to the corresponding feed axes inversely as an offset of the nominal position value. Thus the structural model-based correction represents the complete causal functional chain—from the generation of thermo-elastic errors to their correction (Fig. 16.1) (Großmann 2012). Other authors also map the overall thermo-elastic functional chain for correction purposes (Denkena and Scharschmidt 2009; Brecher and Wissmann 2009; Scharschmidt 2011), but by means of behaviour-oriented equivalent representations, in which a reference to structure is either limited or not possible at all. Conversely, in the structural model-based paradigm outlined in B07, the reference to the machine tool structure, as well as the direct model’s and parameter’s reference to the thermal FE models, is generally maintained.

The approach to the structural model-based correction is given for the example of the “MAX” test bed. The basic conceptual design of the test bed was first described in (Großmann 2009) and is explained in detail in Chap. 1.

16.2 Approach

The modules of the structural model-based correction were drafted according to the thermo-elastic functional chain (Fig. 16.2). The modules’ functionality is briefly elucidated below.

Mechanical and electrical load data are recorded drive related in the *load data capturing* module. The load data may have to be compressed due to data transfer bandwidth limitations at the communication interface between control core and model. Data compression can be done by, for instance, averaging (*load data compression* module). The load variables are consolidated into data records (*create*

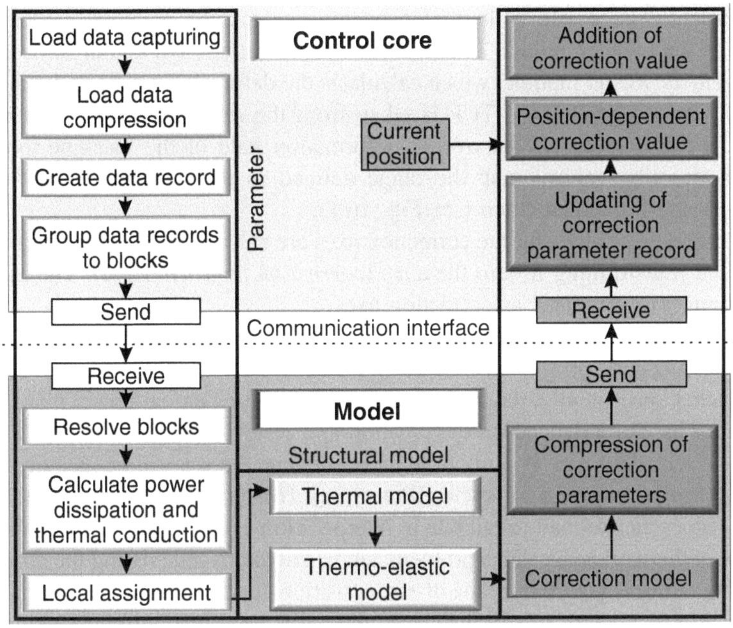

Fig. 16.2 Modules for structure based correction of thermal deformations

data record module). If the sampling interval at which the data records are captured is less than the transmission time of the communication interface, then the data records have to be grouped into blocks before transmission (*group data records to blocks* module). For the models, the blocks have to be resolved again (*resolve blocks* module).

The load-dependent power dissipation (heat sources) and thermal conduction (heat transfers) values of the *thermal model* are determined from the captured load data in the *calculate power dissipation and thermal conduction* module.

Assemblies with mutual relative movement require a *local assignment* module for the allocation of power dissipation and thermal conduction values in the model. Thus, for instance, it is necessary to assign the power dissipation generated for a ball screw in contact with a spindle-nut to the spindle range travelled.

The modules described up to now can be found in the thermo-elastic functional chain under the *parameter* range shown in Figs. 16.1 and 16.2.

The assigned power dissipation and thermal conduction values are transmitted to the *thermal model* module. This module calculates the temperature field of the machine tool. The ambient state, that is the machine tool's ambient temperature, is captured by temperature sensors. This ambient temperature is considered as a constraint when calculating the temperature field. It has to be calculated in thermal real time. For this reason, no detailed FE machine tool model can be employed, and the order of the FE model has to be reduced [FEM-MOR (Großmann et al. 2012a)].

An alternative is to use a thermal capacity-conductance network model (node model) as the reduced model. The calculated temperature field is transmitted to the *thermo-elastic model* module, which calculates the deformation field of the machine tool. The error vector at the TCP is taken from the structure deformation vector, which, in turn, represents the overall deformation field of the machine tool. The modules described above map the range defined as the *structural model* of the thermo-elastic functional chain (see Fig. 16.1).

The correction values for the correction axes are calculated by inverting the error vector and transforming it onto the axes (*correction model* module). The machine tool driving axes are used as correction axes.

The thermo-elastic model is position-specific. The mutual arrangement of the assemblies depends on the current position of the Tool Center Point (TCP) in the workspace. Consequently, the stiffness matrix K and the thermal power matrix K_{WT}, (compare Fig. 16.1) depend on the position and result in position-specific thermo-elastic models. As a rule, it is impossible to calculate these position-specific models in the control's interpolation cycle (IPO cycle). However, it is necessary that correction values can be made available in interpolation cycle as a function of the TCP position in the workspace. This problem can be solved by calculating the correction value in a slower cycle, by means of each position-specific thermo-elastic model at TCP positions regularly distributed over the workspace (supporting positions, see Fig. 16.7). Afterwards, the correction values are compressed by regression (*compression of correction parameters* module) and transmitted to the control (*updating of correction parameter record* module). The *position-dependent correction value* is calculated in a module in the control by means of the current position and the compressed correction values. This method is described in detail in the section "Position-dependent calculation of the correction value". Finally, the correction values have to be added to the nominal position values of the correction axes; this process is represented by the *addition of correction value* module.

16.3 Results

16.3.1 Real Time Thermal Model

The reduced-order thermal model (FEM-MOR) of the "MAX" test bed was developed in parallel with subproject B07 in A05. It was created in Matlab and was not available in its entirety at the beginning of research in B07. For test purposes, the thermal node models of the ball screw-based z drives of the test bed were also generated in Matlab with the goal of calculating the thermo-elastic errors on the axis z as well as φ_x and φ_y afterwards. The alternative usability of node models and reduced-order thermal models was demonstrated in a generalised way in (Großmann et al. 2012b). The node model (see Fig. 16.3, *left*) is designed to provide the same characteristic parameters both for the input and output interfaces as those used for the reduced-order thermo-elastic model from A05 (see Fig. 16.3, *right*), compare

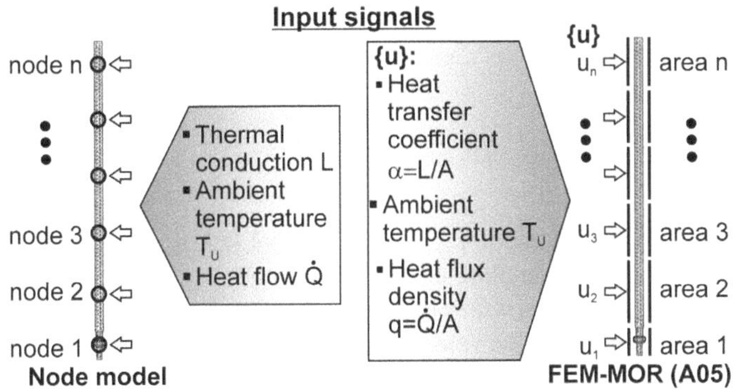

Fig. 16.3 Ball screw spindle as node model and as FEM-MOR

(Großmann et al. 2012a). For this reason, it is possible, in principle, to mutually exchange the two model variants in the *thermal model* and *thermo-elastic model* modules.

16.3.2 Requirements in Terms of the Load Data's Sampling Intervals

The load parameters required as source data to determine power dissipation and thermal conductions are, as a rule, motor current I or motor torque M, velocity v and position x (or the associated rotatory parameters) of the drive axes. The necessary sampling interval can be derived from the limits of jerk j_{max}, acceleration a_{max} and velocity v_{max} of the drives (see follow-up description). These limits were defined in the control during the axes commissioning. The time curves are charted in Fig. 16.4.

The time values Δt_I, Δt_v and Δt_x (see Fig. 16.4 and Table 16.1) are yielded by the maximal rises of acceleration, velocity and position. The current can be assumed to be proportional to driving axis acceleration. Consequently, the sampling interval is determined by means of the acceleration curve. It takes the drive the time value Δt_I

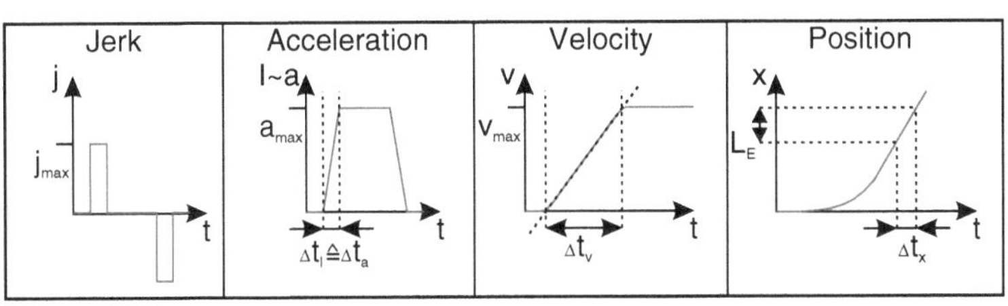

Fig. 16.4 Rise times based on the maximum values of jerk, acceleration and velocity

Table 16.1 Necessary sampling interval for current, velocity and position

	Current	Velocity	Position
Chosen sampling interval	$\Delta t_I = \frac{a_{\max}}{j_{\max}}$	$\Delta t_v = \frac{v_{\max}}{a_{\max}}$	$\Delta t_x = \frac{L_E}{v_{\max}}$
	$T_I \approx \frac{\Delta t_I}{10}$	$T_v \approx \frac{\Delta t_v}{10}$	$T_x \approx \frac{\Delta t_x}{10}$

to achieve maximal acceleration a_{max} with maximal jerk j_{max}. For sufficiently high resolution of current, the sampling interval T_I was chosen, according to the Nyquist-Shannon sampling theorem, with one tenth of Δt_I.

The sampling interval T_v for velocity can be defined analogously to the sampling interval for current.

The position has to be scanned at a sufficient frequency in order to assign it unambiguously to an element in the thermal model. The position sampling interval has to be defined accordingly. Thus we obtain the maximally permitted sampling interval Δt_x from edge length L_E of the elements used in the thermal model and maximal feed speed v_{max}. Thus, in the case of a ball screw, edge length L_E is equal to the axial length of an element. It also makes sense in this case to choose a smaller sampling interval T_x.

From a software engineering perspective, it makes sense to scan all load data in the control core or of the drives at the same sampling interval T_{Core}. For this reason, the lowest calculated sampling interval for all load data (T_I, T_v, T_x) and all driving axes is chosen.

16.3.3 Local Assignment

If the assemblies move relative to one another, *local assignment* of power dissipation and heat transfer or thermal conduction values to nodes and the coupling coordinates in the model is necessary. The *local assignment* approach is explained in the following for the ball screw on the "MAX" test bed. It is necessary to assign the power dissipation generated in the spindle-nut contact that is caused by friction to the coupling coordinates of the thermal model, which maps the spindle. Thermal conduction between the spindle elements and the nut also has to be determined by means of the position (see Fig. 16.5).

As a rule, the integration time increment t_{total}, with which the *thermal model* is calculated and in which the parameters used are constant, is greater than the sampling interval T_{Core} of the load data. Hence power dissipation and thermal conduction calculated over the time interval t_{total} have to be averaged. A solution algorithm in this regard is depicted in Fig. 16.5 for the ball screw. The power dissipation values, determined by means of *local assignment* and affecting the individual elements, are added together and averaged over the time interval $t_{total} = i \cdot T_{Core}$.

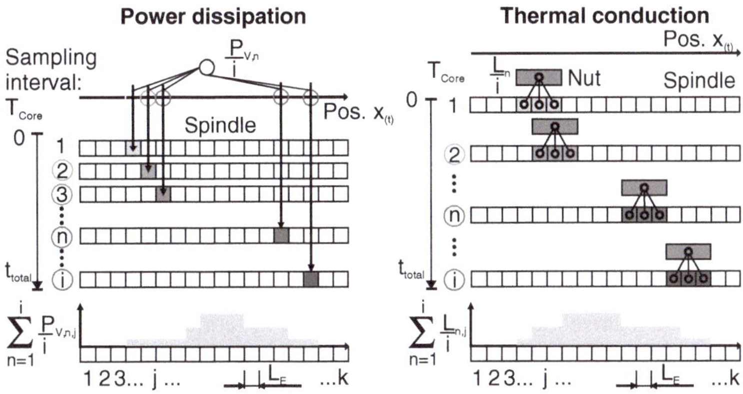

Fig. 16.5 Local assignment of power dissipation and thermal conduction to elements of the ball screw spindle in the simulation model. The elements have an edge length of L_E. Power dissipation and thermal conduction are averaged over the period t_{total}

Regarding characteristics, the *thermal model* is similar to a PT1 element with time constant τ in control engineering and thus functions as a low-pass filter. Since averaging also serves as a low-pass filter, the simulation result is rarely impaired when averaging over one tenth of the time constant. For this reason averaging is permissible.

For illustration, Fig. 16.6 shows the spectrum analysis of white noise filtered both by a PT1 element and a PT1 element with prior averaging. As can be observed in the amplitude spectrum, averaging over a tenth of the time constant rarely influences the frequency spectrum transmitted. The time constant of 1794 s used in the example is the typical time constant for the spindle of the ball screw used in the "MAX" test bed.

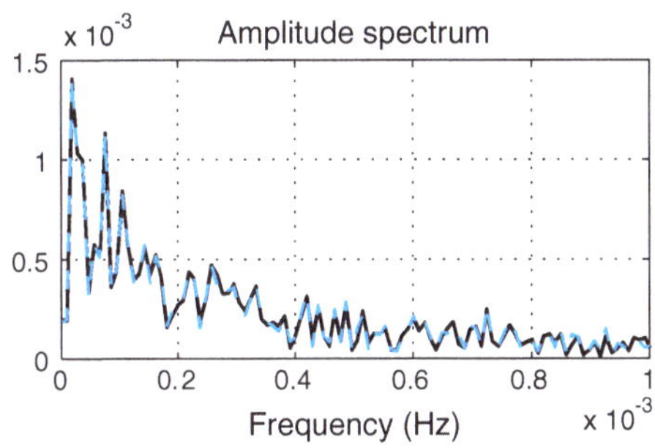

Fig. 16.6 Amplitude spectrum of white noise filtered by PT1 element (*solid line*) and filtered by PT1 element with prior averaging (*dashed line*). The time constant τ of the PT1 element is 1794 s and the average is calculated over τ/10

16.3.4 Position-Dependent Calculation of the Correction Value

The thermally induced change in machine tool deformation is characterized by slowness (Großmann et al. 2013). This means that relevant changes in deformations appear in the range of several minutes (due to large thermal time constants of the assemblies relevant for deformation). In the case of quick mutual movement of the assemblies, however, the resultant error can change at the TCP within milliseconds due to quickly changing transformation of the thermal assembly deformation at the TCP due to the relative movement. For this reason, the correction values have to be calculated position-dependently in the interpolation cycle of the control (IPO cycle, for example 2 ms). This requirement can be met by generating and calculating the thermo-elastic model for supporting positions that are regularly distributed across the workspace (positions of the TCP). The error vector can be interpolated at the TCP (Δw_{TCP}) (see Fig. 16.7, *left*) between these supporting positions. When using the 27 supporting points sketched ($\{\Delta w_{\mathrm{TCP}}\}_{0,0,0}$ to $\{\Delta w_{\mathrm{TCP}}\}_{2,2,2}$), the error vector can be approximated by parabolic polynomials (Großmann et al. 2011):

$$\begin{aligned}\Delta w_{TCP} &= \begin{Bmatrix} \Delta x_{TCP}(x,y,z) \\ \Delta y_{TCP}(x,y,z) \\ \Delta z_{TCP}(x,y,z) \end{Bmatrix} \\ &= \begin{Bmatrix} (a_0 + a_1 x + a_2 x^2)\cdot(b_0 + b_1 y + b_2 y^2)\cdot(c_0 + c_1 z + c_2 z^2) \\ (d_0 + d_1 x + d_2 x^2)\cdot(e_0 + e_1 y + e_2 y^2)\cdot(f_0 + f_1 z + f_2 z^2) \\ (g_0 + g_1 x + g_2 x^2)\cdot(h_0 + h_1 y + h_2 y^2)\cdot(i_0 + i_1 z + i_2 z^2) \end{Bmatrix}\end{aligned} \tag{16.1}$$

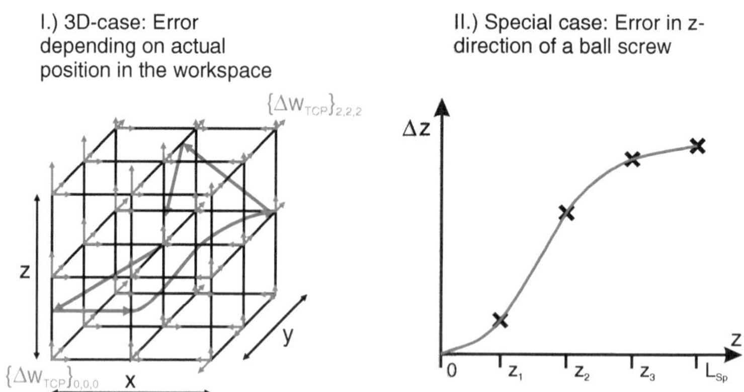

Fig. 16.7 Calculation of the correction value depending on actual position in the workspace (*left side*) or depending on the nut position on the ball screw spindle (*right side*). L_{Sp} is the length of the ball screw spindle

This strategy can be followed analogously in the one-dimensional case for the ball screw in the test bed described above (see Fig. 16.7, *right*). To approach the error in the z direction, for instance, a polynomial can be employed to interpolate the four supporting points shown in the figure:

$$\Delta z(z) = a_1 z + a_2 z^2 + a_3 z^3 + a_4 z^4 \tag{16.2}$$

The polynomial coefficients are transmitted to the control. The error vector can be determined with minimal computation effort in the IPO cycle by means of the polynomial functions and the current position in the workspace.

16.3.5 Implementation of Load Data Capture in the TwinCAT3 Control

Load data capture was implemented for the control Beckhoff TwinCAT3 used for the "MAX" test bed (see Fig. 16.8). A communication program gains access to the data in the control via the ADS interface[1] (ADS-DLL for C/C ++) made available in TwinCAT3. The load data (x, v, I) required are cyclically scanned in the control and transmitted to the communication program by notification. In the program, the data are grouped into blocks and stored together with a time stamp in a shared memory used both by the control core and the thermal model run under Matlab. The shared memory takes on the function of the *communication interface* (see Fig. 16.2). Matlab accesses the shared memory in a time increment given by a timer. Memory access is performed using a mex function written in C. Correct transfer of the data block is checked in Matlab based on the time stamp. Checks are run to determine whether the data block has already been read out or if a data block has been lost. Apart from reading out the load data, the timer also controls the writing of correction values in the shared memory. Furthermore the timer calls the simulation model and ensures that it is calculated in the specified time increment. Thus the timer allows the real time requirements to be fulfilled by the model. The communication program cyclically reads out the correction values in turn using a timer and sends them to the control via ADS interface.

For the communication interface introduced, the model part (here: Matlab) is independent of the specific control. Instead of Matlab, also other programming environments can also be used to implement the simulation model. As a precondition, access to the shared memory and a sufficiently accurate timer should be available.

[1] ADS-Automation Device Specification.

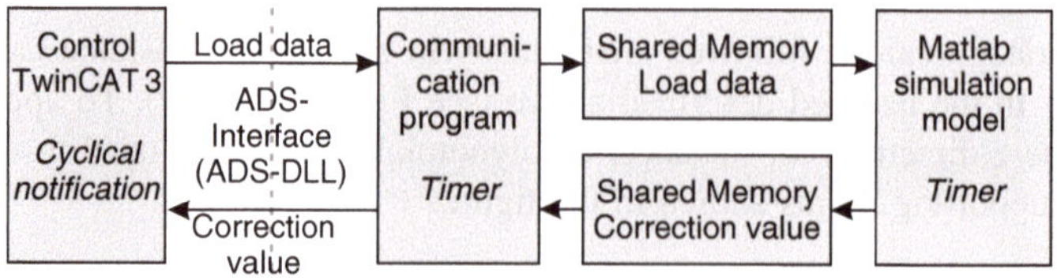

Fig. 16.8 Load data capture and transfer of the correction value

16.3.6 Test of Control-Integrated Load Data Capture and Temperature Field Calculation

Load data capture, communication and integration of models from the subproject A05 were first tested for a mobile demonstrator (MiniHex). The captured load data and the resultant error Δz are depicted in Fig. 16.9. The models comprise the reduced-order thermal FE model of a ball screw of the demonstrator, as well as the thermo-elastic model and the power dissipation and thermal conduction models, see Chap. 7. The models were implemented in Matlab, and the TwinCAT3 was used as control.

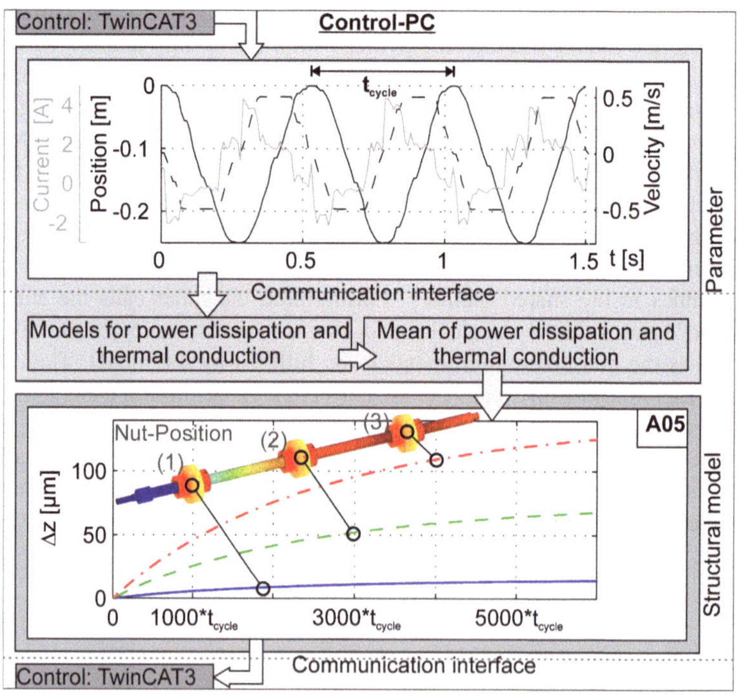

Fig. 16.9 Load data (*gray* current, *dashed line* velocity, *solid line* position) and the resulting error values (Δz) for different nut positions on the ball screw spindle

The load data (I, v, x), as well as the ambient temperature were captured in the control at a sampling interval of 10 ms and grouped into blocks of 10 s. These blocks were transmitted to the Matlab model by means of the communication interface described (see Fig. 16.8). In Matlab, power dissipation and thermal conduction were calculated and averaged. The temperature field was determined by means of thermal conduction and power dissipation values, and the resultant error Δz in the axis direction of the ball screw was ascertained from the temperature field.

16.4 Classification of Outcomes in the CRC/TR 96

B07 provides one of the three correction paradigms explored within the CRC/TR 96. By using a *thermal and a thermo-elastic model*, the approach maps the thermo-elastic functional chain so that the machine tool structure remains visible in the model. The approach to the structural model-based correction is based on algorithms, technologies and models for real-time simulation of the thermo-elastic behaviour (among others, model order reduction methods from the subprojects A05 and A06). Related to this paper, subproject A05 provided reduced-order thermal network models of the test bed.

The parameters of the machine tool for the frictional and contact positions of the guidances and bearings in the thermal network model are set in coordination with subproject B04.

The setting for power dissipation of the driving motor models in the network model of the test bed was parameterised for the machine tool based on motor power dissipation models from C04.

Making available the "MAX" test bed, subproject C06 provides the experimental basis and the control functionality in order to test, verify and demonstrate the correction system. From the correction system, in turn, arise requirements to be fulfilled by the control interfaces as described above, for instance for load data capture.

In parallel, the correction system was tested on a mobile demonstrator (hexapod MiniHex, compare Chap. 7), in which six variable-length struts driven by ball screws move an end effector. The mobile demonstrator is—as was the "MAX" test bed—run with a Beckhoff TwinCAT3 control.

16.4.1 Applicability Options for Structural Model-Based Correction

The focus of the application of structural model-based correction is on machine tools with greatly varying machining processes or ambient conditions. In these cases, the qualitative characteristics of the corresponding deformation-relevant temperature fields are also subject to high variability. Correlative correction approaches (Wu et al. 2006, 2011) map a part of the thermo-elastic functional chain by means of regression, fuzzy or neuronal models and mediate between the measured temperatures and the deformations. In terms of the variability mentioned, these correction approaches require a large number of temperature measuring points to capture the variable temperature fields relevant for deformation at sufficient resolution. Due to the metrological efforts necessary or lack of temperature sensor accessibility to the desired temperature measuring points, this correlative approach may be subject to limitations.

As a precondition for structural model-based correction, access to the load data used in the machine tool control is needed, and it should be possible to add the correction values to the nominal values in position control of the feed axes. During the design and testing of the concrete structural model-based correction, the developer needs thermal modelling know how and sufficiently assured knowledge of all essential power dissipation and heat transfer parameters.

The method provides an opportunity to correct thermally induced errors without additional energy consumption by the machine tool, for instance by additional cooling. Furthermore, except for the ambient temperature sensors that make available the signals involved as a constraint, the application of additional sensors to the machine tool is not mandatory.

16.5 Outlook

The modules labelled "Parameter" and "Structural model" in Fig. 16.2 have been implemented as examples up to now. To be able to demonstrate and test the full correction system, the modules framed at the right of Fig. 16.2 will be implemented for the "MAX" test bed and the mobile demonstrator (MiniHex).

References

Brecher C, Wissmann A (2009) Modelling of thermal behavior of a milling machine due to spindle load. In: Proceedings of the 12th CIRP conference on modeling of machining operations, vol 2, San Sebastian, 7–8 May 2009

Denkena B, Scharschmidt K-H (2009) Modellbasierte Temperaturkompensation für Werkzeugmaschinen. ZWF 104(9):698–702

Großmann K (2009) MAX – Versuchsträger für eine Hochgeschwindigkeits-Leichtbau-Genauigkeitsmaschine. In: Großmann K (ed) Lineardirektantriebe in Werkzeugmaschinen, Dresden, 2009

Großmann K (2012) Thermo-Energetische Gestaltung von Werkzeugmaschinen. ZWF 107 (5):307–314

Großmann K, Jungnickel G, Kauschinger B, Mühl A, Rehn S (2011) Prozessaktuelle strukturmodellbasierte Korrektur thermoelastischer Fehler. In: Conference proceedings of the 1. colloquium of the CRC/Transregio96, Dresden, 28–29 Nov 2011

Großmann K, Galant A, Mühl A (2012a) Effiziente Simulation durch Modellordnungsreduktion. ZWF 107(6):457–461
Großmann K, Städel C, Galant A, Mühl A (2012b) Berechnung von Temperaturfeldern an Werkzeugmaschinen. ZWF 107(6):452–456
Großmann K, Städel C, Mühl A (2013) Simulative Erweiterung der Datenbasis zur korrelativen Korrektur thermoelastischer Verformungen. In: Großmann K (ed) Tradition und Gegenwart bei der Analyse des thermischen Verhaltens spanender Werkzeugmaschinen, Dresden, 2013
Scharschmidt K-H (2011) Model-based method for compensation of thermal deformations of machine tools. Dissertation, IFW Leibniz Universität Hannover
Wu H, Li G, Shi D (2006) Fuzzy logic thermal error compensation for computer numerical control noncircular turning system. In: International conference on automation, robotics and computer vision, Singapore, 5–8 Dec 2006
Wu C-W, Tang C-H, Chang C-F, Shiao Y-S (2011) Thermal error compensation method for machine center. Int J Adv Manuf Technol 59:681–689

Chapter 17
Modelling and Design of Systems for Active Control of Temperature Distribution in Frame Subassemblies

Welf-Guntram Drossel, André Bucht and Christoph Ohsenbrügge

Abstract This paper outlines theoretical and experimental investigations into opportunities for active control of heat flows in frame structures. The application of phase-change materials using metal foam as matrix material makes it possible to achieve an increased thermal capacity that is coupled with the frame structures as needed by means of variable thermal conductance based on active materials.

17.1 Introduction

The more inhomogeneous the occurring temperature fields, the more challenging it is to correct or compensate for thermally induced deformations, particularly if the frame structures do not reach a thermal equilibrium. In practice, the current desire for shut-down measures in order to increase power-efficiency results, however, in an increase in heat flows that are inhomogeneous both in time and location. To counteract the negative consequences of this phenomenon, an intervention must occur in the thermal effect chain at the level of the heat flow and the temperature fields that result from it. Influencing heat flows and temperature fields at the component level can thus enhance the thermal behaviour of the machine as a whole.

W.-G. Drossel · A. Bucht
Fraunhofer Institute for Machine Tools and Forming Technology (IWU), Chemnitz, Germany
e-mail: Welf-Guntram.Drossel@iwu.fraunhofer.de

A. Bucht
e-mail: Andre.Bucht@iwu.fraunhofer.de

C. Ohsenbrügge (✉)
Professorship for Adaptronics and Lightweight Design, TU Chemnitz, Chemnitz, Germany
e-mail: Christoph.Ohsenbruegge@iwu.fraunhofer.de

K. Großmann (ed.), *Thermo-energetic Design of Machine Tools*, Lecture Notes in Production Engineering, DOI 10.1007/978-3-319-12625-8_17

17.2 Approach

Stabilisation given heat inputs that are inhomogeneous in time can be achieved by means of additional thermal capacities and a concomitant higher thermal inertia in the system. In view of this, a thermal equilibrium—once achieved—can be sustained longer against a varying heat flow. However, an increased thermal capacity has a disadvantageous influence on transient behaviour outside the steady state. Increased inertia can greatly delay or even prevent the achievement of thermal equilibrium (Fig. 17.1b). Possibilities for active control of heat flows were considered to improve this transient behaviour. In this process, a high-capacity heat storage structure is to be isolated from the component during warm-up. After the steady state is established, the thermal capacity is raised through thermal coupling with the storage and thus it is possible to input or output heat flows varying in time and/or amount in a controlled manner (Fig. 17.1c).

To implement the heat storage, phase-change materials (PCM) able to absorb or release large amounts of thermal energy within a small temperature range due to the enthalpy of thermodynamic changes of state were investigated. These materials provide high storage density in the range of phase transition (typically from solid to liquid), resulting in a thermal stabilisation in the range of the PCM's melting temperature. To optimise the thermal transfer inside or into the storage, the phase-change material was infiltrated into a substrate made of aluminium foam. Two material-based mechanisms were considered as a way to achieve variable coupling of the storage device to the component to be thermally regulated: first, the anisotropic behaviour of magneto-rheological fluids under the influence of the magnetic field, and, second, the mechanical response of thermal shape memory alloys to changes of the thermal field.

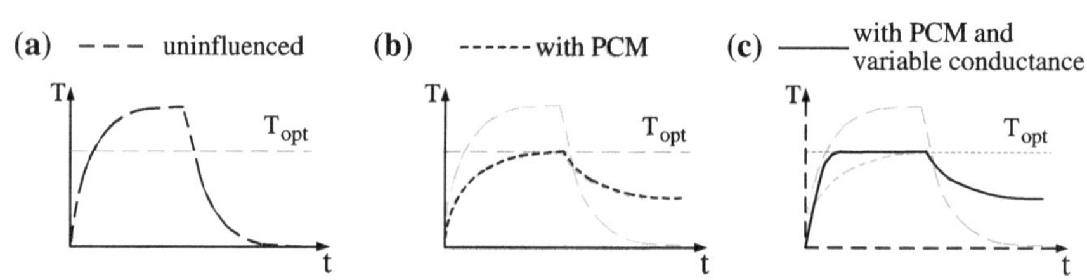

Fig. 17.1 Schematic view showing the influence of the considered measures **a** temperature curve, uninfluenced, **b** with heat storage, **c** with heat storage and variable conductance

17.3 Results

17.3.1 Material Composite of Phase-Change Material and Metal Foam

Closed-cell aluminium foam was considered as matrix for the phase-change material. Encapsulating the PCM in closed cavities ensures that the volumetric expansion in the phase transition (of up to 16 % in case of the PCM based on paraffin investigated here) does not affect the outside. The distribution of the PCM in small amounts in the composite material provides good heat introduction and thus makes it possible to absorb or release large heat flows. The infiltration of the foam takes place at temperatures above the paraffin's melting point after having evacuated the metal foam body or through dipping into a tank with liquid PCM. In this process, micro-cracks in the cell walls are used to enable the pores located inside to be filled (Fig. 17.2).

Fill rates of up to 100 % can be obtained with this technique. However, the actual fill rate depends on the statistical distribution of pores and micro-cracks. The volume of embedded PCM can be increased by a cooling down procedure during infiltration. The volume reduction during PCM solidification causes a partial vacuum, and additional material is drawn in. However, an increase in the fill rate due to this method is only desirable up to complete infiltration in the liquid state. If the cavities are completely filled with already partially solidified PCM, the volumetric expansion in the melting procedure will expand the PCM to a volume larger than the cavities, which is to be avoided.

Designing the matrix as a sandwich structure, in which the cover layers made of solid material are metallurgically bonded to the foamed core, is a measure to improve heat entry into the heat-storage structure. When metal foam structures are used in a sandwich design, two outer sides have already been sealed by means of the cover layers (see also Fig. 17.3). Additional sealing measures are necessary for the open outer sides that remain. In this context, sealing by forming proved to be an efficient method. Here, parts of the foamed core are removed and protruding cover sheet plates are bent and used for sealing. The bent cover sheet plates are either welded or glued.

A homogenized material model representing a substitute material, whose thermal behaviour conforms to that of the material composite, was developed. It is possible to determine the thermally relevant material properties of density ρ_h and specific

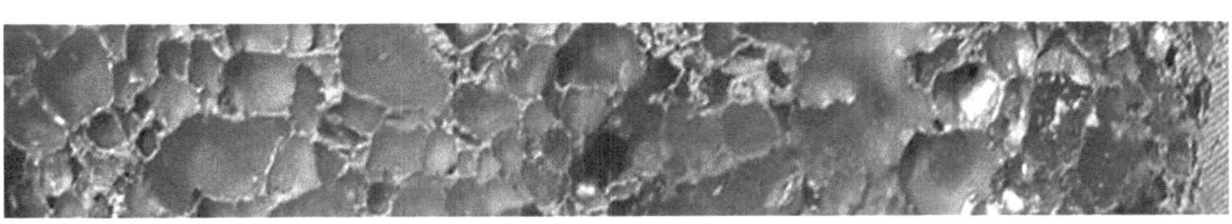

Fig. 17.2 Metal foam with PCM embedded in the pores

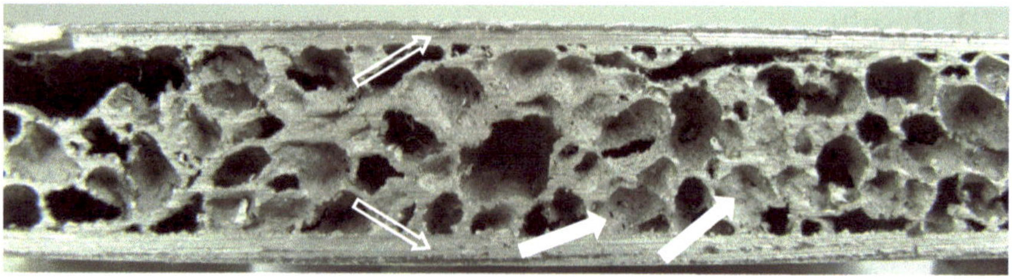

Fig. 17.3 Metal foam structure in sandwich design, micro-cracks in cell walls (*white arrows*), cover layers made of aluminium and metallurgically bonded (*white hollow arrows*)

thermal capacity $c_{p,h}$ for the homogenized material from the material characteristics of the aluminium- and PCM portions of the composite material by using (17.1) and (17.2).

$$c_{p,h} = \frac{m_{alu} \cdot c_{p,alu} + m_{pcm} \cdot c_{p,pcm}}{m_{alu} + m_{pcm}} \tag{17.1}$$

$$p_h = \frac{m_{alu} + m_{pcm}}{V_{ges}} \tag{17.2}$$

The influence of the latent heat that is absorbed in the phase transition can be represented by two approaches: first, by an increased specific thermal capacity in the melting region and, second, by specifying the temperature-dependent enthalpy. From a numerical point of view, the second approach is to be preferred, since the material characteristics do not change abruptly. Thermal conductivity of the metal foam can be determined using the modelling approaches by Singh and Kasana (2004), as well as Bhattacharya et al. (2002).

The two relevant variables—absorbable heat and thermal conductivity—conflict in terms of objectives. Thermal conductivity diminishes as a function of increasing porosity of the metal foam, on the one hand. On the other hand, the PCM volume that can be introduced, and, consequently, the absorbable heat, increase. As a result, it is necessary to weigh the respective importance of the variables based on the given requirements.

A basic demonstrator showed that the thermal behaviour of a plate structure can be influenced by infiltrating it with PCM. The basic body of 0.0013 m^3 volume and 786 kg/m^3 density was infiltrated with 0.49 kg PCM RT 33 by the firm Rubitherm and subjected to a heat flow of 19 W (Fig. 17.4 left). The influence of thermal stabilisation can also be demonstrated by simulation, by comparing metal foam elements with and without PCM (Fig. 17.4 right). Hence the results show the suitability of the approach and are generally in accord with the results previously published by Aggogeri et al. (2010).

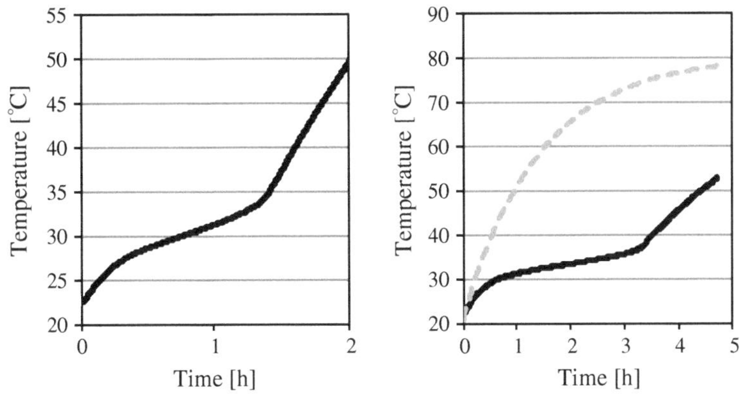

Fig. 17.4 Temperature curves obtained by experiment (*left*) and simulation for metal foam sandwiches with PCM (*dashed line* unfilled, *drawn through line* filled)

17.3.2 Switchable Thermal Conduction Based on Shape Memory Alloys

The working mechanism investigated to influence heat flows by means of thermally active shape memory alloys was to macroscopically influence the contact properties by opening and closing air gaps. Thermal shape memory alloys (SMA) can exert a force after reaching an activation temperature based on the one-way shape memory effect. SMAs exhibit a very large volume specific work capacity in comparison to other actuator-principles. This allows for a lightweight and highly integrated solution. Activation of SMA-actuators can be done actively by heating or by using the heat of the process itself.

Thermal equivalent networks are used to model the thermal conductivity properties. The thermal resistances in a contact are influenced by actuators based on shape memory alloys. Consideration of a simple heat transfer path in a beam structure, which can be influenced by actuators based on the shape memory effect, produced the networks depicted in Fig. 17.5. The heat bridge created in the experiment (Fig. 17.5a) consists of a heat source, a heat sink, and a structure with good thermal conductivity to connect sink and source. A v-shaped section of this structure can be inserted and pulled off via a shape memory actuator working against a spring. Thermal equivalent circuits for the setup are shown in Fig. 17.5b–d. In case of active triggering of the shape memory actuator, it was switched between arrangements with good and restricted heat transfer. In this setup, the shape memory actuator works against a resetting spring element made of conventional material. The latter opens up the transfer path in the basic state because the spring force of the resetting element exceeds the force of the actuator element. Due to heating of the active element, the force exerted increases beyond the reactive component, and the thermal contact is established. In experiments with

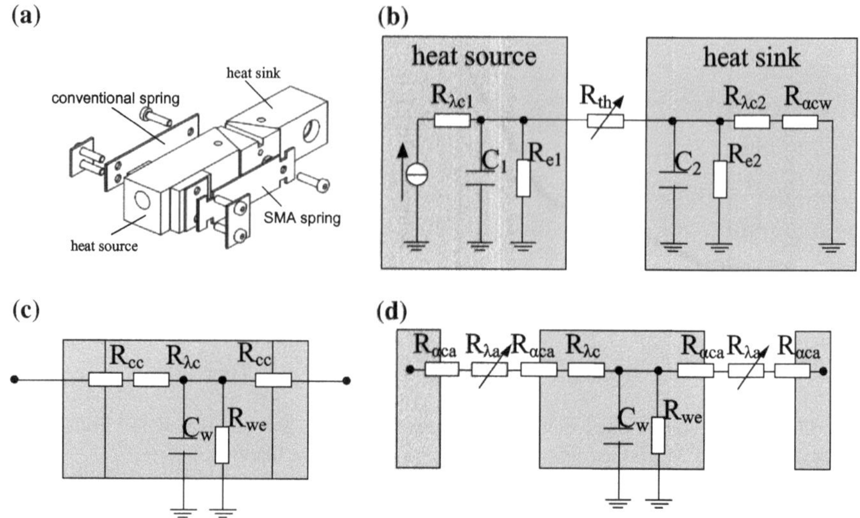

Fig. 17.5 Experimental setup of the heat bridge and thermal equivalent circuits. **a** Overview on overall setup. **b** Network of overall setup. **c** Network for closed heat bridge. **d** Network for opened heat bridge

the path open, heat source and heat sink differed in temperature by 80 K. Upon activation of the shape memory element, the gradient was reduced to 12 K. The resistance in this setup dropped from 16.2 K/W down to 2.3 K/W.

Actuators made of thermal shape memory alloys allow for a self-sufficient mode of operation due to the use of available thermal energy. By contacting an actuator element directly to the heat source, it is activated once a temperature threshold value has been exceeded. Figure 17.6 demonstrates the transient thermal behaviour of a setup like this: First, a high thermal load (30 W) was applied for 3,600 s. In the first case (heat source isolated from heat sink; "open gap"), this resulted in a rapid

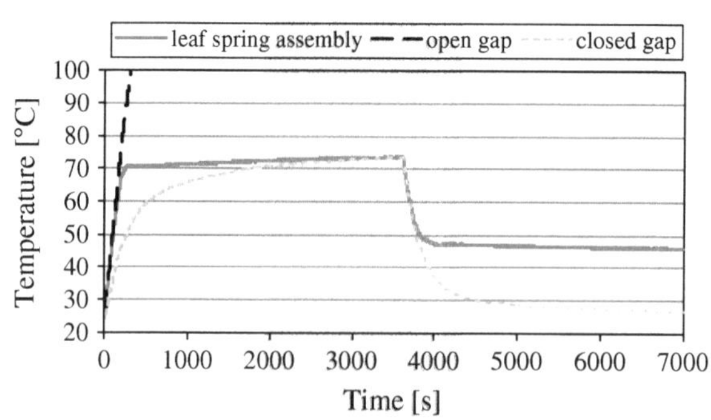

Fig. 17.6 Transient thermal heat source behaviour in experimental setup for an autonomously switching heat bridge. A step change in thermal input power is the input variable

warming up of the structure. With a permanently closed transfer path ("closed gap"), a stationary state was achieved after a delay. Direct coupling of the shape memory element to the heat source, in turn, led to rapid warming up due to the isolation of both source and sink. Reaching the activation temperature in the actuator material closed the air gap, and heat can dissipate into the sink. After reducing the heat supply to 0.7 W, as a consequence, in this case autonomously influenced, the temperature quickly dropped back to the initial state. In the permanently closed case, the final temperature dropped more dramatically (Neugebauer et al. 2012).

17.3.3 Switchable Thermal Conduction Based on Magnetorheological Fluids

Magnetorheological fluids (*MR fluids*) are suspensions of ferromagnetic particles in a carrier fluid that demonstrate significant changes in properties under the influence of an external magnetic field. Whereas most prior investigations and applications of these suspensions mainly focussed on the mechanical properties that are influenced, here the thermal characteristic parameters were considered as a function of the magnetic field. When generating a magnetic field, the magnetic particles of the fluid form chain- or column-like structures. As a result, an anisotropic heat-conducting structure of increased thermal conductivity k_{II} forms in the direction of the magnetic field. Heat transfer is significantly lower if the particles in the fluid are sedimented, for instance, due to the influence of gravity, thus forming two separate layers.

Theoretical preliminary studies by Bruggeman (1935) permit a prediction of thermal conductivity in the uninfluenced state. Reinecke et al. (2008) describe a model for a fluid influenced by a magnetic field. This thermal conductivity depends on the volume content of particles in the overall fluid Φ and the thermal conductivity of the materials used.

Intensity of the magnetic field does not directly affect the model but it can be considered by describing the packing fraction inside the particle chains that increases with field intensity. The model is structured in two stages and, first, the thermal conductivity in the chains k_{chain} under the influence of the magnetic field is calculated by using (17.3). This thermal conductivity is then used in a parallel connection to the surrounding fluid according to (17.4) to determine the effective overall conductivity of the fluid k_{II}. The volume fraction of particles inside the chains Φ_{int} is used as a variable parameter for model adjustment. Extending this approach, it is also possible to predict heat transfer in a deposited fluid: first, a value k_{sed} is determined for the deposited layer with particles with a fraction Φ_{int}, and this is evaluated in comparison with the pure fluid layer in series connection according to (17.5).

$$\left(\frac{k_{chain} - k_p}{k_{base} - k_p}\right)\left(\frac{k_{base}}{k_{chain}}\right)^{1/3} = 1 - \phi_{\text{int}} \tag{17.3}$$

$$k_{II} = k_{base}\left(1 - \frac{\phi}{\phi_{\text{int}}}\right) + k_{chain}\left(\frac{\phi}{\phi_{\text{int}}}\right) \tag{17.4}$$

$$k_{II} = \left(\frac{1}{k_{base}}\left(1 - \frac{\phi}{\phi_{\text{int}}}\right) + \frac{1}{k_{sed}}\left(\frac{\phi}{\phi_{\text{int}}}\right)\right)^{-1} \tag{17.5}$$

An experimental setup was designed to determine thermal conductivity. In the setup, the temperature difference between a heat source and sink was measured in thermal equilibrium under a constant heat supply both with and without a magnetic field. Investigating several samples made it possible to show the functional principle of increased thermal conductivity due to the influence of the magnetic field. The results also outlined the influence of the particle volume fraction. However, in the case of very high volume fractions, the difference between the two switching states is diminished. This can also be explained by the model, since the volume fraction in the chains or deposited layers is limited by the possible packing fraction. If the total volume fraction of particles in the suspension is equal to the maximal packing fraction, then the suspension's thermal conductivity can no longer be influenced because no variation of particle density can be achieved. In Fig. 17.7, the predictions by the model are compared with the values measured for spherical iron particles in silicone oil. The values influenced by the magnetic field were measured at a magnetic field intensity of 105 mT, while the deposited state was reached after a latent period of at least 12 h.

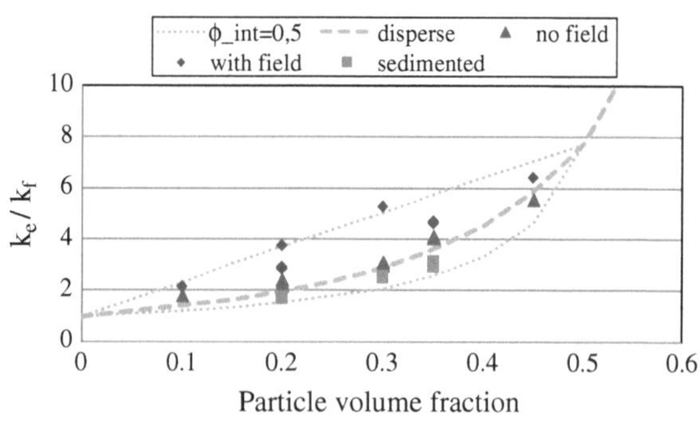

Fig. 17.7 Thermal conductivity as a function of the particle volume fraction, related to thermal conductivity of the basic fluid. Theoretical (*dashed curve* for disperse case, *dotted curves* for magnetic field influence and deposited fluid at an internal volumetric fraction of 0.5) and measured values (*check marks*) for thermal conductivity, scaled to the basic fluid's thermal conductivity

17.4 Classification in the CRC/TR 96

The thermal behaviour of individual components and parts can be actively or passively influenced by the materials and the composite materials investigated, and thus, the overall thermal behaviour of a machine tool can be enhanced. More and more partially heterogeneous thermal behaviour is to be expected due to increasing lightweight construction, as well as anticipated measures towards more energy-efficient utilisation of machine tools, such as switching off subsystems and units during breaks in machining and, as a result, reducing thermal basic load. It is possible to decrease these inhomogeneities by means of the functional mechanisms of compensation investigated here. A lower range of variation in the temperature field reduces the effort required for correction methods to compensate for thermally induced errors in the TCP. The temperature field can be stabilized particularly at positions with a substantial entry of dissipated power and strongly varying heat sources, such as in feed drives, bearings or electronic components.

17.5 Outlook

In future project steps, switchable thermal conduction based on shape memory alloys will first be investigated in more detail. The focus will be on maximizing the switching effect and on an investigation of the behaviour as a control system. Both approaches to switchable thermal conduction have to be evaluated in terms of their applicability according to the requirements in frame structures.

The composite structure consisting of PCM and metal foam will be explored in more detail in co-operation with subproject A07 (see Chap. 9). The diffuse domain method allows for easy discretization of complex geometries. The characteristic values of a metal foam-PCM composite structure determined by means of this method are to be compared both to the simulation results based on the simplified material model and to experimental data.

Finally, the combination of the working principles of thermal storage and variable thermal bonding are to be verified in experiments, and the consequences to the thermal field of a predefined machine structure will be simulated.

References

Aggogeri F et al (2010) Multifunctional structure solutions for ultra high precision (UHP) machine tools. Int J Mach Tools Manuf 50:366–373

Bhattacharya et al (2002) Thermophysical properties of high porosity metal foams. Int J Heat Mass Transf 42:1017–1031

Bruggeman DAG (1935) Berechnung verschiedener physikalischer Konstanten von heterogenen Substanzen. I. Dielektrizitätskonstanten und Leitfähigkeiten der Mischkörper aus isotropen Substanzen. Ann Phys 24(7):636–664

Neugebauer R et al (2012) Control of thermal flow in machine tools using shape memory alloys. WASET Int Sci Index 71:493–498

Reinecke BN et al (2008) On the anisotropic thermal conductivity of magnetorheological suspensions. J Appl Phys 104:023507

Singh R, Kasana HS (2004) Computational aspects of effective thermal conductivity of highly porous metal foams. Appl Therm Eng 24:1841–1849

Chapter 18
Structurally Integrated Sensors

Christian Wenzel and Michel Klatte

Abstract The work described here explores the suitability of metrological applications for direct measurement of thermo-elastic deformations in a machine tool's basic structure with the goal of correcting positioning errors by means of control engineering. This paper first introduces the measurement principle of the sensors applied. It then presents the outcomes of investigations aimed at the system approach's validation, demonstrating the concept's suitability to measure structural deformations. It was also possible to show that the deformation of simple structural elements can be represented with sufficient accuracy by the linear functions of the signals obtained using structurally integrated sensors.

18.1 Introduction

In currently available machine tools, it is common practice to directly measure the thermal expansions or deformations of spindles (Yoshimi et al. 1989) or tool holders (Stahlgren et al. 1996) and to correct them by means of control engineering to avoid workpiece errors. However, only a portion of the overall temperature-induced TCP displacement can be measured by this method, since these components represent only a relatively short path in the kinematic chain between the workpiece and the tool, which is completed by the entire machine structure.

When expanding the approach of uniaxial measurement and correction of local thermo-elastic deformations into a general and comprehensive concept, it is thus necessary to involve the deformation of the whole kinematic chain and thus of the machine structure (Biral et al. 2006; Mitsuishi et al. 2001). Thus it is possible to

C. Wenzel · M. Klatte (✉)
Fraunhofer Institute for Production Technology (IPT), Aachen, Germany
e-mail: michel.klatte@ipt.fraunhofer.de

C. Wenzel
e-mail: christian.wenzel@ipt.fraunhofer.de

K. Großmann (ed.), *Thermo-energetic Design of Machine Tools*,
Lecture Notes in Production Engineering, DOI 10.1007/978-3-319-12625-8_18

implement the correction of the TCP displacement based on direct measurement of the real deformation.

Consequently, the subproject C03 targets the implementation of metrological applications to directly capture thermo-elastic deformations in the machine tool's basic structure by means of an integrated sensor system during part machining. Another objective is to calculate the correction data for the tool centre point (TCP) based on the measured values recorded.

The sensor system, which consists of many individual path measurement sensors, makes it possible to measure the local thermo-elastic relative displacement at relevant measuring points. A geometric-kinematic transmission model of the machine representing the location of the individual machine tool sensors and their kinematic behaviour will be engineered. Using this model, the individual displacements measured are expected to be converted into an overall displacement related to the TCP and added to the nominal position values as correction data.

18.2 Configuration of Sensor Applications

18.2.1 Measurement Principle

The measurement principle employed in capturing the thermo-elastic deformations is based on thermally stable rods integrated into the machine structure (see Fig. 18.1). Each thermally stable rod is mounted in the machine structure without stresses via a fixed and floating bearing arrangement. If the length of the surrounding structure changes, for example, due to thermo-elastic deformation, the structure displaces the thermally stable rods on the floating bearing side. It is possible to directly capture this displacement by means of specific displacement sensors.

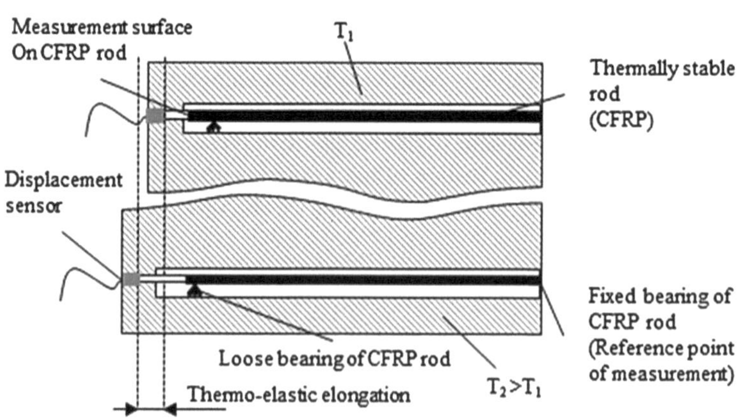

Fig. 18.1 Measurement principle for direct measurement of thermo-elastic deformations

Table 18.1 Material parameters of the applied thermally stable rods made of fibre-reinforced plastics

Property	Value
Thermal coefficient of extension	-0.1×10^{-6} K^{-1} longitudinal to fibre
Fibre orientation	Unidirectional
Fibre volume percentage	Approx. 65 %
Thermal conductivity	17 W/m K
Specific heat	0.71 J/g K

The thermally stable rods are tube-shaped to minimize deflection due to weight. Their thermal coefficient of extension ought to be several orders of magnitudes less than that of steel or other structural materials, so that it can be considered negligible in deformation measurement. For this reason, the rods are made of carbon fibre-reinforced plastics (CFRP) by a pultrusion method, so that all fibres inside have the same orientation longitudinally. The material parameters yielded are presented in Table 18.1.

18.2.2 Sensor Arrangement in Complex Machine Structures

When several sensor applications are arranged in the same area, even complex deformations can be mapped. Bending states for the corresponding structure can also be represented by means of two sensors arranged in parallel. The difference between the signals of two sensor applications can be derived and analysed in this state. This is elucidated in Fig. 18.2.

Thus, bending around two axes can be mapped by means of a three-dimensional arrangement of three sensor applications. In both cases, the elongation of the structure along an axis can also be directly measured.

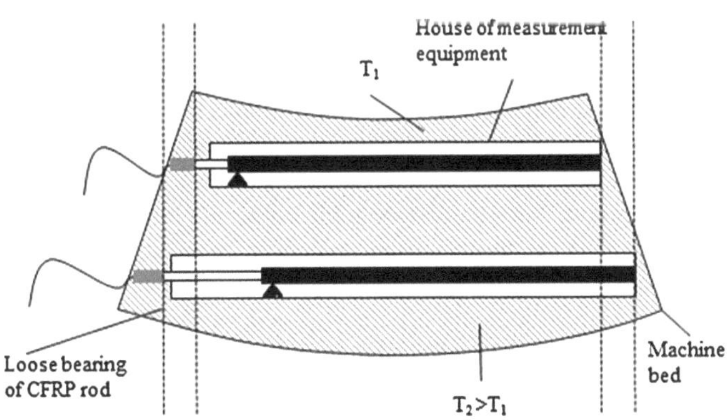

Fig. 18.2 Capture of complex deformation states (in this case: bending) by using several measurement applications—diagrammatic view

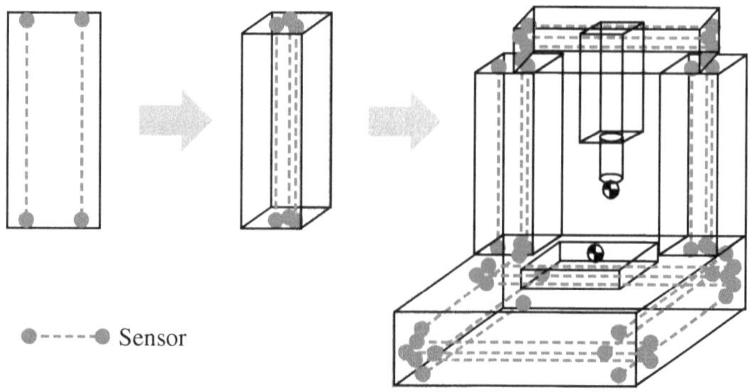

Fig. 18.3 Extending the sensor concept to complex structures, shown for a bridgeport milling machine

As mentioned previously, a sensor application can be employed to directly measure the deformation state of a simple structure, such as a cuboid element. Machine tool structures are frequently relatively simple in design. Thus, for instance, the approximated representation of a machine column could be achieved by a three-dimensional arrangement of three sensor applications. However, if a more complex machine structure must be mapped, or the complete machine, this can be done by linking several sensor groups.

The machine structure is subdivided into cuboid elements (see Fig. 18.3). Each cuboid element includes a sensor arrangement. Connecting several cuboid elements makes possible, for instance, an approximate mapping of the complete structure of a bridgeport milling machine.

18.3 Test Bed Results

18.3.1 Determining the Suitability of the Sensor Applications

Sufficiently high measurement accuracy is one of the requirements that must be fulfilled by the sensor applications. The measurement result should especially not be distorted by stick-slip effects in the bearing elements. The signals have to be stable over the long term, since the sensor applications are functionally permanently integrated into the machine tools and are supposed to make values available for correction by means of control engineering.

Investigations of measurement accuracy and long-term stability tests were conducted in order to verify first of all whether these requirements are fulfilled by the sensor applications introduced in the section.

Tests were also done to determine the measurability of deformation states of simple structural elements by means of a several sensor applications.

18.3.1.1 Verification of the Measurement Principle

Tests Designed to Determine the Suitability of the Sensor Application

Investigations to test the suitability of the sensors in the complete sensor setup for the application were carried out on a simply designed test bench. The test bench setup is explained in the following.

The basic test bench geometry is a cuboid. The cuboid is made of mineral casting which is a typical structural material, and is equipped with two structurally integrated sensor applications (see Fig. 18.4). In these applications, different displacement sensors can be employed for deformation measurement, making it possible to test the usability of different displacement sensors. After concluding a series of tests with different sensor principles, eddy current transducers were ultimately chosen for the demonstrator. For this reason, the following explanations are based on the results gained during the verification of the transducer type. To verify the measurement results of the structurally integrated sensors, external capacitive displacement sensors that measure outside the structure against the same reference surfaces, are used. For these investigations, the test bench is located on an assembly plate and run in a thermally stabilised (±1 °K) environment (Table 18.2).

To induce different deformation states, the test bench is fitted with ten heating cartridges altogether (five per side). The heating cartridges are arranged in parallel so that they can be run independently of each other at different heating power

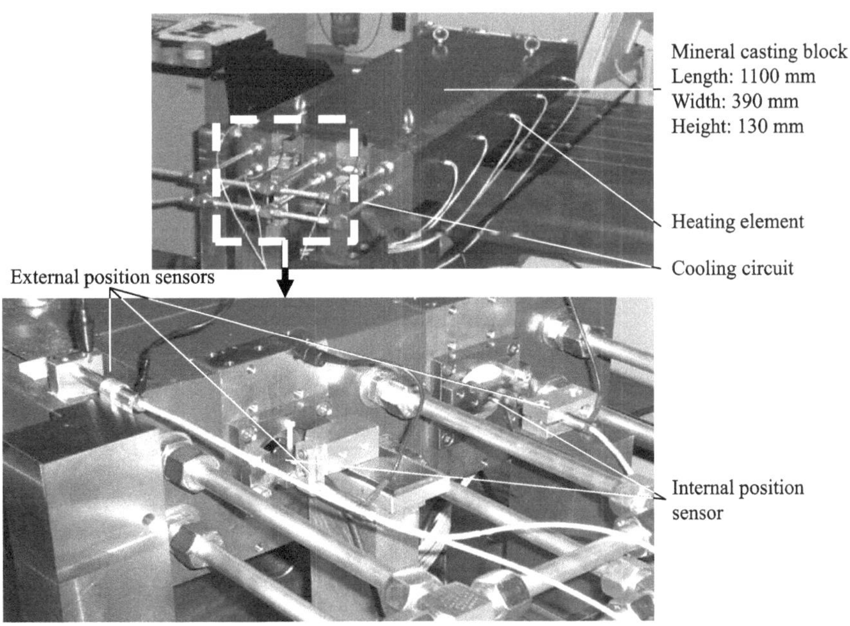

Fig. 18.4 Test bed to verify the measurement principle

Table 18.2 Material parameters of the mineral casting material used

Property	Value (approx.)	At a temperature of (°C)
Thermal coefficient of extension	15×10^{-6} K^{-1}	20
Thermal conductivity	2.9 W/m K	25
Specific heat	0.73 J/g K	25
Thermal diffusivity	1.75 mm^2/s	25

values. This way both symmetric and asymmetric deformation states of the test bench can be generated.

In the first verification measurements, the cuboid was first homogeneously, that is symmetrically, heated by 500 W of power. A uniaxial thermo-elastic elongation state formed. After a measured elongation of approximately 80 μm, the heating was switched off so that the cuboid's elongation retracted. This period is defined as the cooling down phase.

The measurement results provided in Fig. 18.5 show that there is an approximately 2 % maximal relative deviation between the signals of the structurally integrated sensor applications and the signals obtained from the external sensors. This observation is independent of the measurement principle of the displacement sensor used. As intended, it is possible to measure the elongation of the ambient structure with very high accuracy at the point where the sensor application is fixed.

An asymmetric deformation of the structure was initiated in ongoing tests. To do this, the cuboid was heated via the integrated heating cartridges by 250 W of power over 1 h. The two structurally integrated sensors clearly show different deviations according to their own position in the cuboid. Consequently the bending states can be identified using the difference between the two signals (Fig. 18.6).

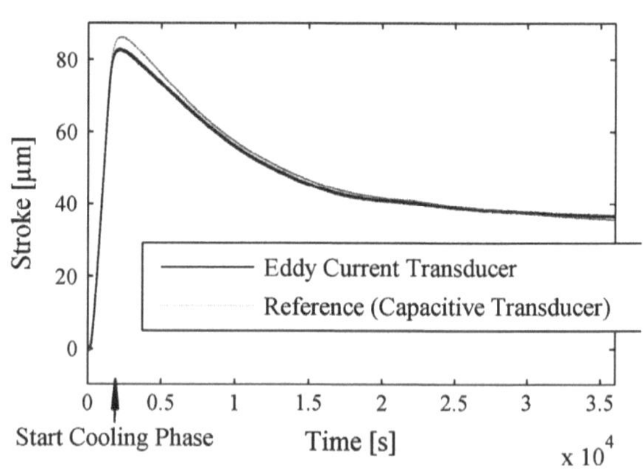

Fig. 18.5 Measurement results obtained for the verification of the sensor application by means of eddy current transducers

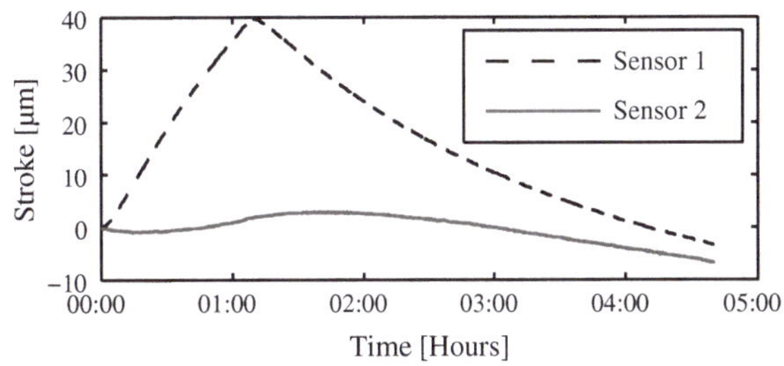

Fig. 18.6 Verification of recording bending by means of two sensor applications—measurement results

18.3.1.2 Long-Term Stability Tests

To guarantee that the measurement results obtained remain valid over a longer time and to ensure that it is not necessary to recalibrate the total system too often, the displacement sensors should have long-term stability. Since eddy current transducers can also be affected—as a condition of the principle according to which they function—due to temperature variations, corresponding investigations were performed.

The goal of these investigations was to exclude to the greatest extent possible all external error factors for the measurement and to track the changes in the measured signal over a longer period. When external influencing factors are excluded, it can be concluded that corresponding instabilities in the measured signals are caused by the sensor itself.

The test bench is mainly composed of a simple sensor holder made of Invar to provide a stable temperature (see Fig. 18.7). This fixture houses three displacement sensors that measure together on a surface (target). The distance from displacement

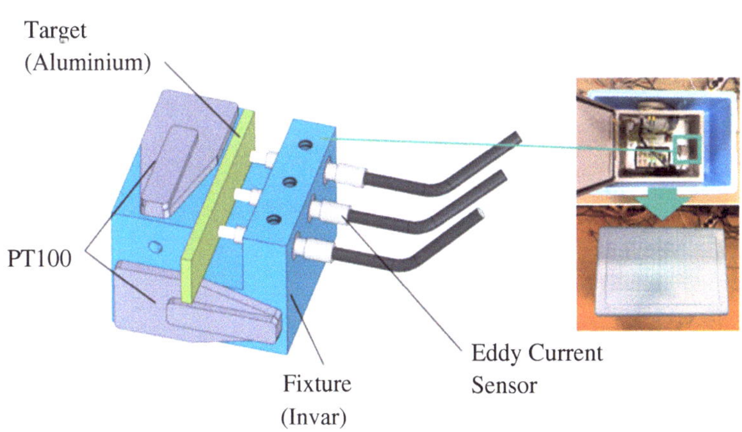

Fig. 18.7 Test setup for long-term stability tests

sensor to target is adjusted to a fixed value. However, the distance between the displacement sensors and the target is influenced by the thermo-elastic characteristics of the sensor fixture. For this reason, the total setup is placed in a thermally isolated reservoir that decreases temperature variations in the environment and ensures that they do not directly affect the temperature of the measurement setup. The total test setup is run in an air-conditioned environment, in which the temperature variations should be less than 1 °C. Nevertheless, the sensor fixture's temperature is captured at two positions and recorded.

Electromagnetic interference also influences the measurements. To minimise the influence from sources of electromagnetic disturbances, the overall setup is run in an insulated environment. Inside this insulated space, there is only the sensors' evaluation electronics.

The stability test was performed as follows. The test bench was established as described above and the sensors were put into operation. Then the recording was started. The measurements of the sensor signals and at the fixed temperature sensors were executed without additional external impacts over more than 7 days.

As a result (see Fig. 18.8) it can be stated that the temperature of the sensor fixture could be kept constant within 1.5 °K during the experiment. This value corresponds to the theoretical error influence on the test signals of (to be calculated: effective length and thermal coefficient of extension).

According to our observations of the first 6 h, all of the sensors drifted in the same direction by approximately 0.8–1.2 μm, as illustrated in the Figure. This drift does not correlate with the setup temperature and thus could result from sensor electronic heating. After this heating stage at the beginning, the signals remain constant within a band of 0.33 μm up to maximum 0.5 μm (sensor 3). Maximal drift is also charted in the diagram in Fig. 18.8.

Furthermore, it was also found that the sensors' temperature dependency is approximately 0.05 μm/°C. Sensor temperature should ideally be kept constant to avoid measurement errors. However, sensitivity to variations in the ambient temperature is relatively low. Consequently, slight temperature variations can be tolerated.

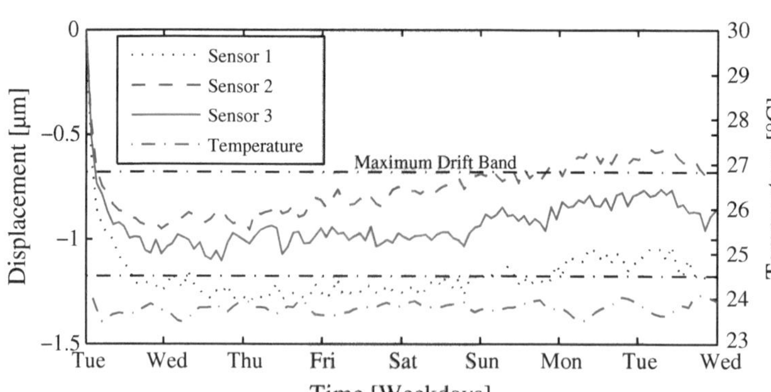

Fig. 18.8 Measurement result of the long-term stability test of eddy current transducers

The dominant influence on temperature dependency results from the sensor electronics. Therefore the sensor system should only be put into service after a sufficiently long warm-up phase for the sensor electronics. The displacement sensors can be regarded as stable under these conditions.

18.3.2 Validation Measurements

This section focusses on the exploration of more complex deformation states, particularly bending. Fundamentally the relationship between the real deformation and the signals from the structurally integrated sensor applications had to be clarified. This relationship forms the basis of the deformation models necessary to implement a correction by means of control engineering.

For validation measurements, the test bench for fundamental investigations introduced in Sect. 3.1 is extended as follows. The cuboid is clamped on one side for the follow-up tests. In the case of asymmetric heating, a bending state should appear which causes the structure to be deflected against the clamp. This type of mounting approximately corresponds to a machine tool column.

A prerequisite for the external capture of more complex deformation states is the ability to determine the elastic line or deflection of the test bench. This can be achieved if a series of displacement sensors measure in the direction of the lateral face of the test cuboid. This setup is shown in the diagram in Fig. 18.9. Ten

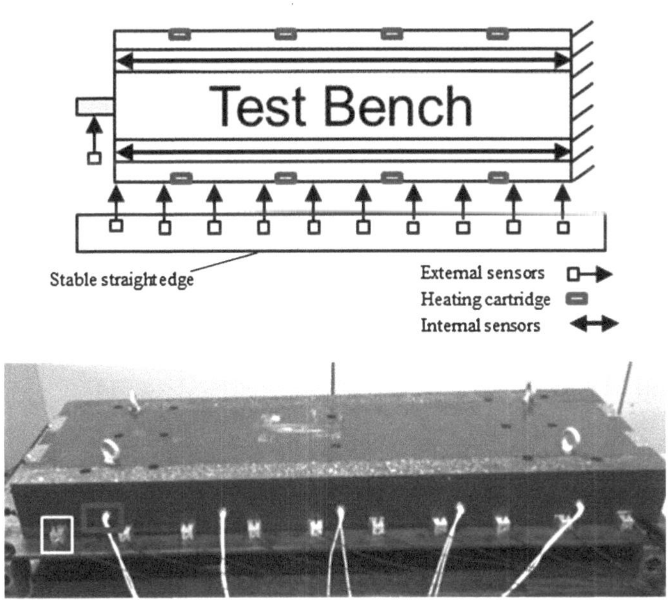

Fig. 18.9 Extension of the test bench for fundamental investigations for validation measurements

displacement sensors are available to determine the elastic line, so that it is possible to measure the elastic line using ten support points at high local resolution at each point in time. A displacement sensor is also employed in order to directly measure the deflection of the unclamped side of the cuboid.

Before each measurement, the cuboid is in a state of equilibrium with the environment. In this state, the cuboid has a homogeneous temperature distribution of approximately 20 °C.

Before switching on the power supply for the heating cartridges, homing of all displacement sensors is performed. Starting from this home position, the thermo-elastic cuboid deformation is observed.

After 1 h of heating, the power supply for the heating cartridges is switched off, and the cuboid cools down slowly until the ambient temperature has been reached. Cooling down was recorded over at least 4 h to make sure that any retraction and possible changes in the thermally induced deformation in the cooling down phase can also be recorded. In all measurements, the measurement values were taken at 1-s intervals.

Thus, (Fig. 18.10), for instance, elucidates the measurement of the elastic line by means of external sensors at five points in time. At the beginning of the test, the values of all sensors were set to zero, which means that the elastic line is understood to be a straight line. The strongly curved lines show the state of maximal deformation, as well as the deformation state at points just before and after the maximum. As is to be expected, a clear bending of test bench and a total displacement or deflection of the left side can be observed. The fifth curve, in turn, is almost a straight line per se, that means bending has returned to the initial state.

To determine whether there is a linear relationship between the deflection and the signals of the integrated sensors, the following approach was applied. First the deflection value was defined as the parameter to be investigated. For modelling, these parameters could be directly used, for instance, in order to describe the characteristic of the machine column deflection.

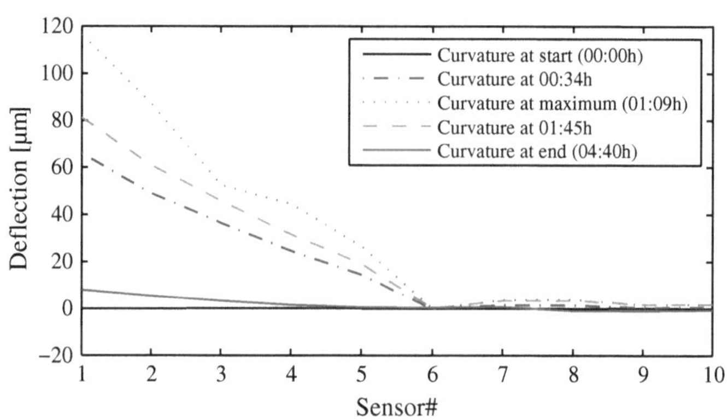

Fig. 18.10 Measurement results of the recorded elastic line at test bench for fundamental investigations

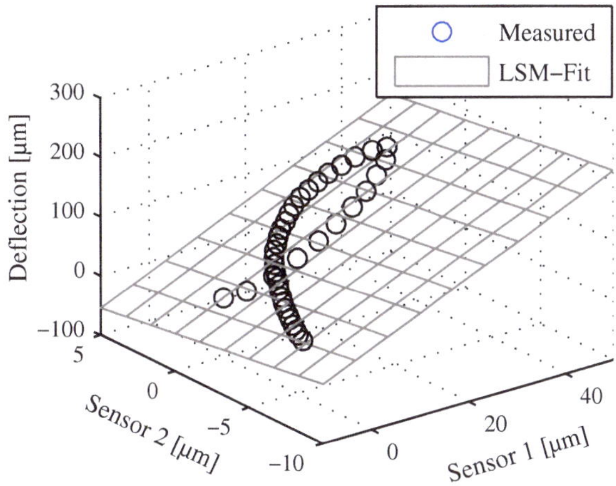

Fig. 18.11 Representation of sensor signals over the measured cuboid deflection with deflection plane

In Fig. 18.11, the signals of the structurally integrated sensors and the deflection parameter are shown. The ascending branch indicates a heterogeneous expansion of the test cuboid at the positions of the individual sensors (sensor 1 shows a higher deflection than sensor 2) and a corresponding increase in thermo-elastic deformation. During the subsequent cooling down period, the points accordingly go back to their initial state (both the internal sensors and the deflection).

This diagram clearly shows that all of the points measured are situated in the vicinity of a plane. This indicates a linear relationship between the externally measured deflection and the signals from the structurally integrated sensors.

The function of this plane could then be found through a compensation algorithm using the least square method. Consequently, the linearity of the relationship is provided by the maximal point—plane distance. Linearity was determined to be (as) 2.16 % of the measurement range (2.98 µm maximal absolute deviation).

The same error also appeared in the temporal representation (see Fig. 18.12). In this diagram, both the measured deflection curve and the deflection curve that was

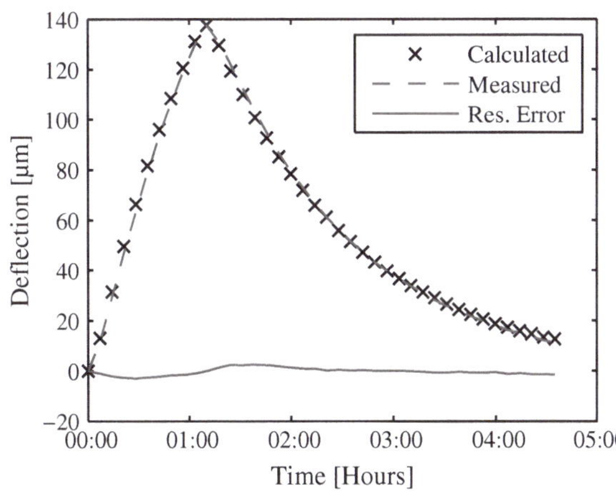

Fig. 18.12 Deflection as a function of time—measured and obtained from compensation calculations—in comparison

calculated across the compensation plane, as well as the uncorrected error were charted. A very high concordance between the graphs and thus high linearity were indicated, especially in the heating phase.

Based on the tests, it can be concluded that signals from structurally integrated sensors can be used to characterise the deformation state of a structural element using simple, linear models for the corresponding load cases.

18.4 Classification in the CRC

The subproject C03 "Structurally integrated sensors" is immediately connected with the overall topic of the CRC and intersects in various ways with other subprojects in the consortium.

Thanks to the structurally integrated sensors, it is possible to immediately capture the thermo-elastic deformation of the machine structure. Structurally integrated sensors offer an additional source of data for any kind of correction methods (subprojects B05–B07). As a result, not only machine data or temperatures can be employed as input variables, but also information about the real thermo-elastic deformation of the machine structure. Furthermore, the approach permits validation of the methods to calculate structural deformations from subproject B07 on the real machine bed.

As a part of project area C, this subproject is another opportunity to enhance the machining accuracy of machine tools by means of design measures in machine frames and components.

18.5 Outlook

In the ongoing project phase, a three-axis bridgeport milling machine will be commissioned with a system consisting of 21 structurally integrated sensor applications. The sensor applications can be verified in connection with a complex geometric structure by means of this demonstrator. Furthermore the deformation models developed in the project can be extended to complex structures and validated with the demonstrator. Thus the basis for a correction of the TCP displacement by means of control engineering will be generated.

References

Biral F (2006) A new direct deformation sensor for active compensation of positioning errors in large milling machines, Istanbul, Türkei, pp 126–131

Mitsuishi M (2001) Development of an intelligent high-speed machining center. CIRP Ann Manuf Technol 50(1):275–280

Stahlgren B, Stubbe R, Ericsson C (1996) Arrangement for sensing elastic deformation in a tool stem in a machine tool
Yoshimi T, Matsuura A, Takashima T, Nakano H (1989) Apparatus for compensating for the thermal displacement of the main spindle of a machine tool

Chapter 19
Thermo-Energetic Motor Optimisation

Stefan Winkler and Ralf Werner

Abstract The thermo-elastic deformations of a machine tool can only be determined if the heat sources inducing them are known. In the electric drives of a machine tool, heating results from the losses that appear during the transformation of electric into mechanical energy. The objective of subproject C04 is to model the individual dissipation mechanisms and the heat fluxes responsible for them. To do this, calculation equations suitable for the mechanisms were created and adapted to the motor types used in the machine tools. The models generated this way were subsequently validated through measurements. Based on the physical interactions among the losses, the authors succeeded in engineering a method by which the dissipations can be actively controlled via motor control. The extent to which this dissipation control can be utilised in order to specifically adapt the heating of the drive for the required warming up of the machine tool is still to be clarified. For thermal modelling, unlike classical modelling, the main focus is not on the internal motor structure, but on heat transfers from the drives that are directed towards the machine tool and/or the environment.

19.1 Introduction

The electric drive dissipations of a machine tool are a part of the machine's most significant heat sources. The target of subproject C04 is to calculate these dissipations and the resulting heat fluxes that contribute to the warming up of the machine tool.

S. Winkler (✉) · R. Werner
Professorship Electrical Energy Conversion Systems and Drives, Faculty of Electrical Engineering and Information Technology, Chemnitz University of Technology, Chemnitz, Germany
e-mail: stefan.winkler@etit.tu-chemnitz.de

R. Werner
e-mail: ralf.werner@hrz.tu-chemnitz.de

K. Großmann (ed.), *Thermo-energetic Design of Machine Tools*,
Lecture Notes in Production Engineering, DOI 10.1007/978-3-319-12625-8_19

For this purpose, first the dissipation models of the individual drives have to be created. The dissipations found this way are used as follow-up input variables for the thermal models, by means of which the desired heat fluxes from the drives to the surrounding parts of the machine tool are ultimately calculated.

To minimise additional motor fabrication costs, the models are structured so that they can function without additional sensors, such as further temperature sensors. As a result, the only available input variables are those that can be obtained via machine tool control.

Another requirement is to design the models as simply as possible so that they can be calculated in real time by the control. This ensures that the temperature distribution caused by the drives at any point in time is available for correction methods within the machine tool.

19.2 Approach

The first step is to determine the relevant motor types that are used as drives in machine tools. This selection is a prerequisite for the generation of the power dissipation models.

When engineering the dissipation models, in general, attention must be paid to the subdivision of the overall dissipations into load-dependent and load-independent dissipations. The special features of each motor have to be taken into account. The design of synchronous motors, for instance, is without rotor windings so that dissipation due to current heat only appears in the stator.

As soon as the dissipation models have been established, they are verified by measurements on selected motors. In the next step, the thermal models are generated for the relevant models using the dissipation as input variables.

The primary focus in thermal modelling is on the drives' interaction with the other machine tool components. In this process, the motor design type and the way in which it is attached are decisive factors, since these issues determine the direct contact between the motor and the machine tool frame. The type of cooling and the coolant used are also important for the distribution of the heat fluxes.

The heat fluxes inside the motor play a subordinate role in heat exchange with the other machine tool components. For this reason, thermal modelling can be performed without exact modelling of the internal motor structure.

19.3 Results

19.3.1 Power Dissipation Models

In general, advanced machine tools are equipped with two different motor types. Permanent magnet synchronous motors are preferentially employed as feed drives due to their high power density, which allows them to be designed compactly.

Asynchronous motors are mechanically robust and easy to control. As a result, they are frequently used as work spindle and accessory drives. Other motor types are not applied in machine tools. Consequently, modelling only considers the special features of asynchronous and synchronous motors.

Basically, the power dissipation models can be adapted to all motors, since all motors are characterised by the same loss types. Consequently, an analogous strategy is chosen to determine the losses of all motors. According to their principle, they can be subdivided into four categories.

Heat dissipation from current: These losses appear in all motor windings due to current conduction.

Cyclic magnetisation losses: These losses are caused by the change of the magnetic field within the ferromagnetic motor components. They are also called magnetic or hysteresis losses $P_{V,Fe}$.

Frictional losses: Frictional losses can be grouped into two types—losses generated by bearing friction and losses due to air friction and ventilator.

Additional losses: Additional losses involve all losses that do not fit into the categories mentioned above. These miscellaneous losses occur for different reasons, for instance skin effects in conductors, reluctance losses due to the changing location of the stator and rotor teeth with regard to each other or by harmonic waves in the magnetic field. They are strongly dependent on geometry and magnetising of each motor. All of them are alike—in opposition to the other loss types—in not being directly measurable. Thus, they can be estimated at best.

The stator-side Joule's heat losses $P_{V,W,S}$ are calculated in the same way for all motors (DIN 2008). The results are obtained by means of the equation

$$P_{V,W,S} = 1.5 \cdot R_{Kl} \cdot I_L^2 \tag{19.1}$$

Variables to be inserted are resistance R_{Kl} between two stator terminals and the effective value of conductor current I_L. Terminal resistance can either be determined by means of resistance measurement or be taken from the motor's data sheet. For exact quantification of losses during the motor run, the temperature dependence has to be considered. The effective current value can either be measured autonomously by means of a separate measuring device, or the value provided by the control can be employed. When doing this, it must be taken into consideration that the measuring devices of the frequency converter used for the control can have a measurement uncertainty of max. 5 %.

In asynchronous motors, Joule's heat losses also appear in the rotor. To provide a robust motor design, in general, short-circuit cages are in use. However, with these cages, neither rotor current nor resistance can be measured. Based on the model for the asynchronous motors, the losses can be calculated with the following equation

$$P_{Mech} = P_\delta - P_{V,W,R} = P_\delta - s \cdot P_\delta \tag{19.2}$$

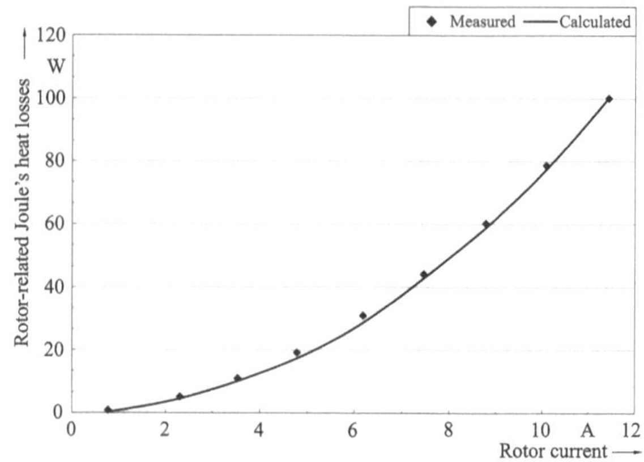

Fig. 19.1 Measured and computed rotor losses of an induction motor with slip rings in comparison

Various approaches, e.g. in Nürnberg and Hanitsch (2001), exist to calculate the air gap power P_δ, necessary to compute the rotor-related Joule's heat losses considering different aspects of power flow inside the motor (Fig. 19.1).

To clarify which of the paradigms provides the most accurate result, comparative measurements were conducted with slip ring induction motors, since both rotor current and resistance can be directly measured on these motors. The most exact consistency is yielded if the air gap power is computed as following

$$P_\delta = P_{El} - P_{V,W,S} - P_{V.U} \tag{19.3}$$

Cyclic magnetisation losses $P_{V,U}$ depend on the adjusted rotary frequency and the motors' magnetisation. In case of asynchronous motors, magnetisation is determined by the amount of applied voltage, whereas, in synchronous motors, it is fixed by the magnets. In principle, the cyclic magnetisation losses are measured in a no-load measurement. In case of asynchronous motors, a fixed rotary frequency is set up, and voltage varies between the half and full nominal voltage. Then the cyclic magnetisation losses together with the frictional losses $P_{V,R}$ yield the load-independent losses:

$$P_{V,U} + P_{V,R} = P_{El} - P_{V,W,S} \tag{19.4}$$

The load-independent losses can be subdivided using the iron loss resistance R_{Fe} as a function of voltage U

$$P_{V,U} + \mathrm{P_{V,R}} = \frac{\mathrm{U}^2}{\mathrm{R_{Fe}}} + \mathrm{P_{V,R}} \tag{19.5}$$

Iron loss resistance is specified in the data sheet as part of the extended equivalent circuit diagram. When using the value from the data sheet, it has to be taken into consideration that the value can have a deviation of up to 20 % due to variations of the material values and manufacturing tolerances.

Synchronous motors have a constant magnetic field since they use permanent magnets. It is possible to vary the magnetic field by impressing an additional magnetising current via motor control. However, this measure increases the motor's required idle power and thus its losses. Furthermore, there is the risk that the magnets may be permanently demagnetised if the field is weakened too much, making the motor unusable. For this reason, an adequate determination and division of losses, as is common practice for asynchronous motors, is impossible. Dismounting the rotor would be necessary for a division of the load-independent losses.

It is also possible to use the motor's S1 characteristic by means of the data sheet. This characteristic specifies the motor's operating point S1 in continuous operation, in which the stator windings achieve a certain final excess temperature, usually 100 K, in comparison with the environment. Since the synchronous motors carried out as servo-design do not have ventilators, their capacity to emit heat into the environment is the same at each operating point. Since, at a standstill, it is only possible to convert load-dependent losses, the total losses $P_{V,Ges}$ can be determined by means of the motor constant as Joule's heat losses—$P_{V,W,S}$ and in terms of additional losses $P_{V,Z}$. Afterwards, it is possible to calculate the load-independent losses as a function of speed n

$$P_{V,U}(n) + P_{V,R}(n) = P_{V,Ges} - P_{V,W,S}(M) - P_{V,Z}(M) \tag{19.6}$$

for each other point of the S1 characteristic by means of the motor constant. Then it is possible to approach the load-independent losses obtained thus from the data sheet or no-load measurement for the loss model as a speed-dependent function.

As previously mentioned, the frictional losses are subdivided into two types—bearing friction and gas friction. Bearing friction is provided using the frictional coefficient, the bearing load, and the rotational speed. Gas friction is proportional to the friction-producing surface and the cube of the rotational speed. Calculation of friction mainly depends on which bearing is used with which lubricant, or, for instance, the consistency of the rotor surface. Since, in general, these variables are as little known as bearing pre-stress, which has a significant effect on the bearing load, each determination of frictional losses by calculation can only provide an estimate in the best case. An accurate determination of friction can only be gained by measurements. If no measurement is carried out, for estimation, the following equation from electrical engineering (Müller and Ponick 2009) provides a solution:

$$P_{V,R} = k_{V,R} \cdot D \cdot \left(l + \frac{0.8^3 \cdot 0.6 \cdot D \cdot \pi}{2p} \right) \cdot (\pi \cdot n \cdot D)^2 \tag{19.7}$$

as well as rotor diameter D, rotor length l, number of pairs of poles p *and* speed n (Fig. 19.2).

Finally the additional losses should be mentioned. These losses can only be determined indirectly by comparing the absorbed and emitted motor power. The method is in principle highly uncertain for measurement, so the additional losses

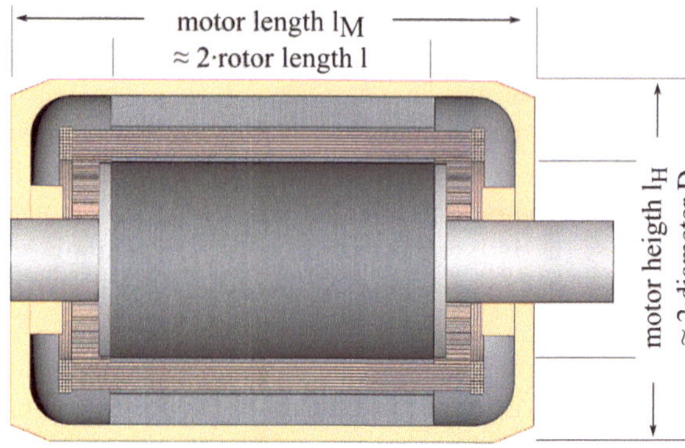

Fig. 19.2 Drive dimensions

measured are accordingly inexact. Calculation of the additional losses can only be estimated. As been demonstrated by the motors measured, an assumption of

$$P_{V,Z} = 0.1 \cdot P_{V,W,S} \tag{19.8}$$

supplied the best conformity with the given measured results (Fig. 19.3).

This calculation equation can be employed for all motors, independently of motor type and nominal power. The pulse frequency and the pulse pattern of the converter control have just as minimal a measurable influence on the additional losses.

With knowledge of the losses and their dependencies from the drive parameters, the drives can be operated at any point of operation, which is optimal in terms of losses. This is enabled by a variation of magnetic flow Φ.

Magnetic flow directly affects the cyclic magnetisation losses and provides the load current required and thus indirectly the Joule's heat losses according to the formula

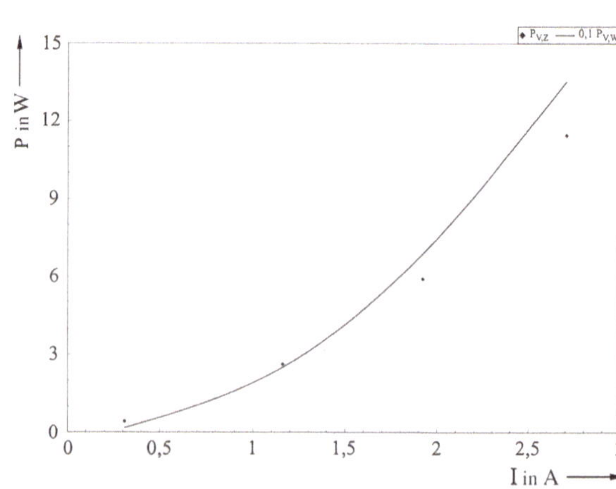

Fig. 19.3 Measured and calculated additional losses, absorbed on an asynchronous machine, in comparison

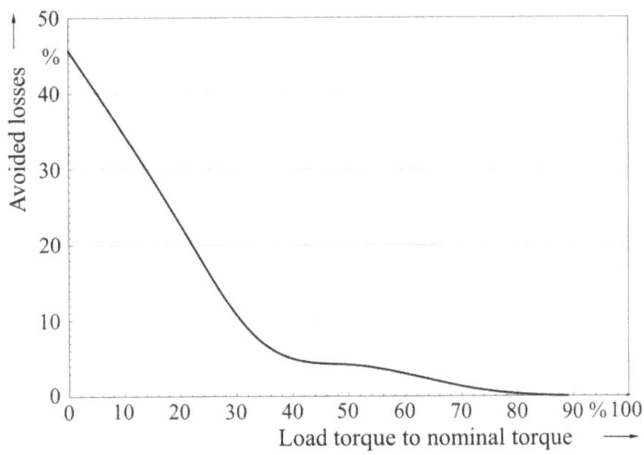

Fig. 19.4 Potential saving in losses due to changed magnetic flow

$$M = k \cdot \phi \cdot I \tag{19.9}$$

According to the present state of the art, magnetic flow is kept constant, if possible, so that the drive can load the maximal torque at any time. The approach followed by the subproject anticipates adapting magnetic flow as a function of the operating point. Thus, for instance, in the partial load range, losses of up to 10 % can be avoided. The no-load losses can even be reduced by almost half (Fig. 19.4).

19.3.2 Thermal Models

The thermal models for the drives must fulfil two functions. On the one hand, they are intended to calculate the temperatures inside the motors, for the temperature-dependent motor parameters, such as winding resistance. On the other hand, they have to enable heat flux to other components of the machine tool to be calculated. For this purpose, subproject C04 employs network models consisting of nodes. An advantage of this is that the structure of a motor can be rebuilt with these models, as by means of an FE model. They can also be converted into electrically equivalent networks and then be computed by using methods of electrical network analysis. The required heat fluxes are provided as currents in the corresponding branches of the network, and the temperatures accordingly as the potentials of the nodes (Fig. 19.5).

A network has to be established for each motor type, in order to take into account the special abovementioned characteristic features of the motors. However, the inner structure is not known, in general. For this reason, a largely simplified standard model can be generated for each motor type at best. The measurements carried out up to now have demonstrated that a standard model like this is sufficient to determine the inner temperature distribution. Concerning the accuracy of the

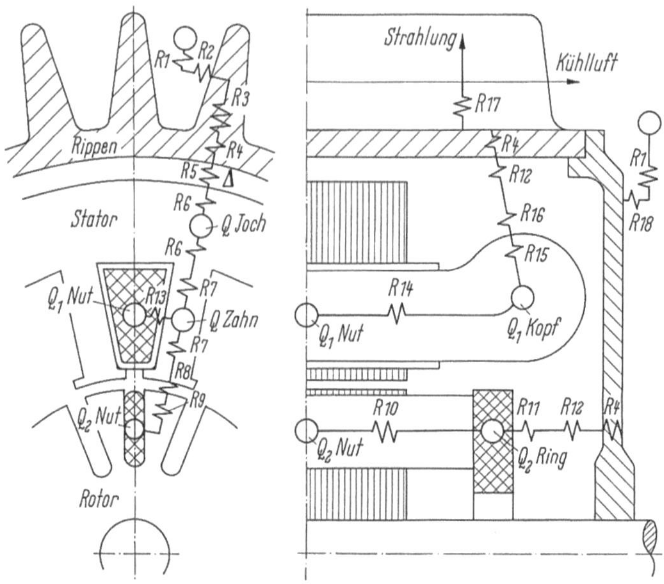

Fig. 19.5 Basic structure of an electrically equivalent heat source network, courtesy (Wiedemann and Kellenberger 1967)

higher ranking machine tool model, the transfers from the motors to the environment are of significantly higher importance than the heat fluxes inside the drives.

Consequently, the correct determination of the heat transfers to coolant, environment and into the machine frame are the major goals in the generation of the thermal models.

19.4 Classification in the CRC/TR 96

Subproject C04 provides methods that are required to specify the losses in the drives of a machine tool. Based on these losses, the thermal models that are created in cooperation with the other subprojects make it possible to calculate heat fluxes between the machine tool, the motors and the environment.

19.5 Outlook

The generation of power dissipation models for the individual motor types has been completed. At present, it remains to be seen to what extent an adaptation of the control that is optimal in terms of losses influences the everyday operation of machine tools.

As they are, the thermal models are too complex for real-time calculation inside the machine tool control. Current investigations are determining the degree to which it is possible to simplify the inner motor structure. The aim is to reduce the number of nodes as much as possible and to model the thermal behaviour with sufficient accuracy. In this context, the linear motors pose another problem. Unlike the compact design of rotating machines, in linear motors, we cannot assume that the whole motor is warmed homogeneously. The main focus of further research is here on the determination of symmetry conditions in order to simplify the model.

References

DIN (2008) DIN VDE 0530 Drehende elektrische Maschinen—Teil 2-1: Standardverfahren zur Bestimmung der Verluste und des Wirkungsgrades aus Prüfungen (ausgenommen Maschinen für Schienen- und Straßenfahrzeuge)

Müller G, Ponick B (2009) Elektrische Maschinen, 1. Grundlagen elektrischer Maschinen. WILEY-VCH, Weinheim

Nürnberg W, Hanitsch R (2001) Die Prüfung elektrischer Maschinen, 7th edn. Springer, Berlin

Wiedemann E, Kellenberger W (1967) Konstruktion elektrischer Maschinen. Springer, Berlin

Chapter 20
Technical and Economic Benchmarking Guideline for the Compensation and Correction of Thermally Induced Machine Tool Errors

Hajo Wiemer, Lars Neidhardt, Werner Esswein and Richard Braun

Abstract The approaches to the correction and/or compensation of thermally caused machining errors developed in the CRC/TR 96 have different effects on the accuracy that can be achieved and on the throughput capacity. Consequently, the approaches generate different benefits. At the same time, the costs and time required for their generation and application differ as well. As a result, not only the technical, but also the economic consequences of each method in terms of its usability for practice-relevant applications are of interest to potential users. Methods for comparing the various approaches are developed in subproject C05. These methods are based on the technical and economic benchmarking guidelines introduced in the paper.

20.1 Introduction

The goal of the research activities is a holistic, comparative assessment or benchmarking of the compensation and correction methods (Fig. 20.1) developed. The term holistic here includes specifying the consequences for the machine tool development process resulting from the solutions, as well as comparing the solutions' suitability for different conditions of use in a high-quality and cost-effective operation. The requirements to be fulfilled by a holistic benchmarking method like this are as follows:

H. Wiemer (✉) · L. Neidhardt
Faculty of Mechanical Engineering, Institute for Machine Tools and Control Engineering, Technical University, Dresden, Germany
e-mail: hajo.wiemer@tu-dresden.de

W. Esswein · R. Braun
Technical University, Dresden, Germany
e-mail: werner.esswein@tu-dresden.de

R. Braun
e-mail: richard.braun@tu-dresden.de

K. Großmann (ed.), *Thermo-energetic Design of Machine Tools*,
Lecture Notes in Production Engineering, DOI 10.1007/978-3-319-12625-8_20

correction method	***compensation method***
• correction based on high-dimensional characteristic diagrams (B05) • correction based on feature model (B06) • correction based on structure model (B07) • correction based on measurement (C06)	• frame assemblies made of smart materials (C02) • frame assemblies with controlled heat flux (C02) • frame assemblies with integrated sensors (C03) • motors optimised ac-cording to thermo-energetic criteria (C04) • fluidic systems opti-mised according to thermo-energetic criteria (A04) • thermally optimised clamping fixture (A01)

Fig. 20.1 Correction and compensation methods, assigned to the subprojects

- The compensation and correction methods, whose application results in different machine configurations, have to be described by their distinguishing features and the parameters relevant for benchmarking.
- The various technological use cases have to be formulated. The operational performance- and market-related constraints have to be involved.
- For benchmarking of methods, evaluation criteria for benefit and cost (such as additional resources needed) have to be formulated.

There is at present a lack of benchmarking methods that combine technical and economic aspects. As a rule, the technical efficacy of innovative solutions is assessed based on experience. The technical benchmarking of machine tools has hitherto been conducted with regard to individual aspects. Abele et al. (2010), Lindner and Götze (2011) and Denkena et al. (2010) proclaim that life cycle costs are suitable for a cost-related balancing of machine tools. The VDMA standard sheet no. 34160 predefines the conditions for their determination (Verband Deutscher Maschinen- und Anlagenbau e.V. 2012). Lindner and Götze (2011) propose to model the components as well as the development and operational processes of the machine tools as a basis for cost determination, for which basic features are lacking. Domain-specific languages adapted for business process modelling, as mentioned in Zor et al. (2011), could provide support.

Strategic models that make it possible to evaluate technological feasibility and profitability at an early point in the development process were developed, among others in Ehrlenspiel (2009) and Eversheim and Schuh (2005). These models were created at the level of development and design methods and of engineering management. However, no immediate relationships to the business analysis of the workflows exist that are connected with the development.

The determination of the machine tool's accuracy is an essential issue for benefit benchmarking from the CRC/TR 96's perspective. There exist sufficiently established largely standardised measurement techniques, such as (DIN 1999), on the one hand. On the other hand, test parts are frequently used, which—however—are not particularly designed to detect thermally induced errors.

Consequently, powerful individual methods are available, but, as a rule, they require detailed data that are not available to the CRC/TR 96. This is mainly information about the machine tool development costs and company-specific organisational sequences connected with the machine tool. These stand-alone approaches are not purposefully integrated into a technical-economic benchmarking method. This means that the methods available do not work in practice.

As a result, the current research is aimed at a suitable benchmarking model mapping the characteristics of the correction and compensation methods and their potential cases of application, as well as the interactions with the business environment, which comprises cost-benefit evaluation criteria for the comparison of the methods. Techniques are also needed that can determine the required information or parameters as efficiently as possible. The model is expected to provide benchmarks first for the methods to be developed within the scope of the CRC/TR 96 in order to define real states and elucidate improvement potential. Finally, the model should also be made available to the machine tool manufacturers and users to provide decision support when implementing the techniques in a machine tool or purchasing a machine tool.

20.2 The Benchmarking Model

The general model developed for benchmarking is shown in Fig. 20.2. The model brings together the required technical and economic levels of consideration and consists of the following partial models: "machine tool configuration", into which the compensation and correction methods are integrated, "technological application cases", "operational constraints" and "market-related constraints". The benchmarking model also includes the "metrics for the valuation criteria" to assess benefit, cost and time, as well as the methods for determination of parameters and benchmarking.

Figure 20.3 introduces the guidelines for the economic comparison of the methods. For benchmarking purposes, the compensation and correction methods

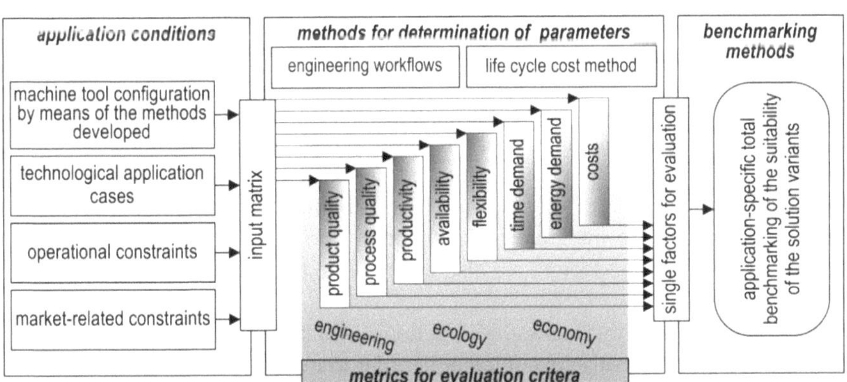

Fig. 20.2 Model approach to assess the methods intended to reduce the thermally induced machine tools' dislocations

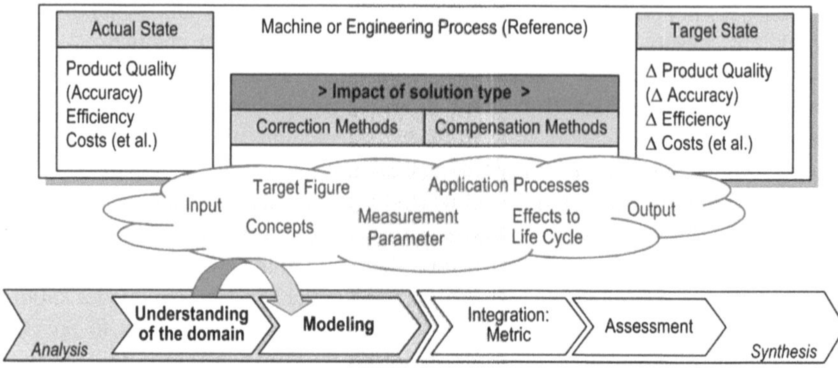

Fig. 20.3 Method assessment framework (Braun and Esswein 2014)

developed must first be brought into a system in a standardised manner, as well as the fundamental concepts to be represented by their input and output variables. It is also necessary to analyse the activities of the subprojects and to document the expected workflows that are affected by the development, implementation and application of the methods. Furthermore, the workflows' effects on the machine life cycle costs have to be estimated. Finally, all of the factors in the evaluation are brought together in a metric for holistic benchmarking. Further details on the framework are described in Braun and Esswein (2014).

The strategy of conceptual modelling (Schütte and Rotthowe 1998) which is common in business informatics, is employed for systematic analysis and representation of the workflows for the development, implementation and use of the CRC/TR 96 methods and connected with the model-based management approach (Esswein et al. 2010).

20.2.1 Partial Model "Machine Tool Configuration"

The "machine tool configuration" partial model describes the machine superstructures that result from a potential application of the different methods in a basic machine. A hierarchic functional structure is chosen for a simple and clearly structured representation of the machine tool configuration. The compensation and correction methods are integrated as a function into the basic machine structure. The benchmark-relevant factors are also entered into the partial model according to function.

The detail level of the hierarchic functional structure is modified according to each task. Thus, two levels are sufficient in order to represent different methods in a basic machine (see Fig. 20.4a). Additional details in the functional structure are needed to benchmark the methods for different basic machine tools that, for example, differ in their frame design type (Fig. 20.4b). Modelling can be supported

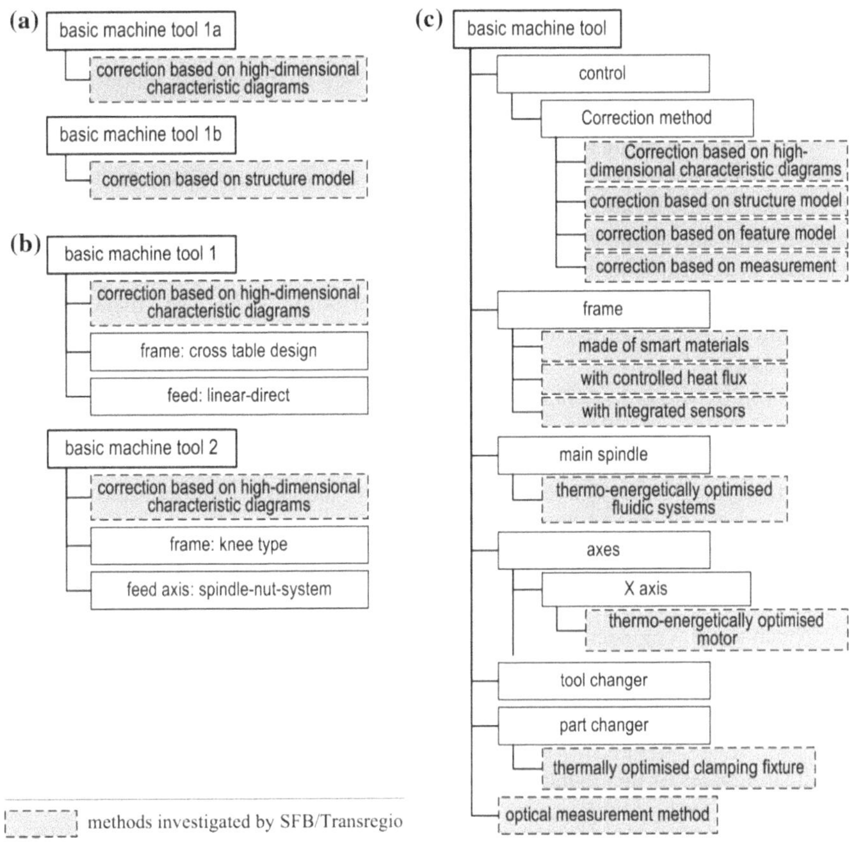

Fig. 20.4 Hierarchic functional model as a basis for the partial model "machine tool configuration" **a** method for a basic machine, **b** method for different basic machines, **c** general classification of the machine tool's functions

by specifying a general system (Fig. 20.4c) with the commonly used functional characteristics and the methods traced. The concrete machine tool configuration is defined from this system by selection of the corresponding functions.

20.2.2 Application Conditions

A machine tool configuration's suitability for the operator's purposes has to be assessed by taking into account the current application conditions. For this reason, the application conditions are shaped by the "technological application cases", "operational constraints" and "market-related constraints" together.

The “technological application cases” involve data for the parts to be machined, such as size, required tool number, lot size and accuracy requirements. It is possible to derive typical machining and setup times, as well as the process output, by using these data. From lot size, it can be determined how often the machine tool has to be reset for another workpiece (classification in Fig. 20.5 according to Schlegel 2002).

The “operational constraints” represent relevant data on production, such as the work time regime or air-conditioning of the factory floor. Run and down times, as well as the restart frequency of the machine tool, result from the work time regime, whereas air conditioning of the factory floor influences the transmission of external thermal influences to the machine tool (Wiemer et al. 2014). The “market-related constraints” consider different cost situations in purchasing, as well as for energy and personnel.

To simplify the considerations of the new methods’ benchmark, first, the same market-related constraints and the same operational constraints are assumed for all operators. Thus, the technological application conditions are the only variable. Three reference cases that are typical for the industry were chosen from the assortment of potential technological application cases (see Fig. 20.5):

- “Diemaking”: workpieces of 5,000 mm max. edge length, 1–5 pieces, high tool number
- “Mechanical engineering”: workpieces of max. 800 mm edge length, part quantity 100, medium number of tools
- “Motor manufacturing”: parts of max. 200 mm edge length, part quantity 10,000, low tool number.

The outcomes of the application condition benchmarking can also be applied by the machine tool manufacturers, since they know under what conditions the

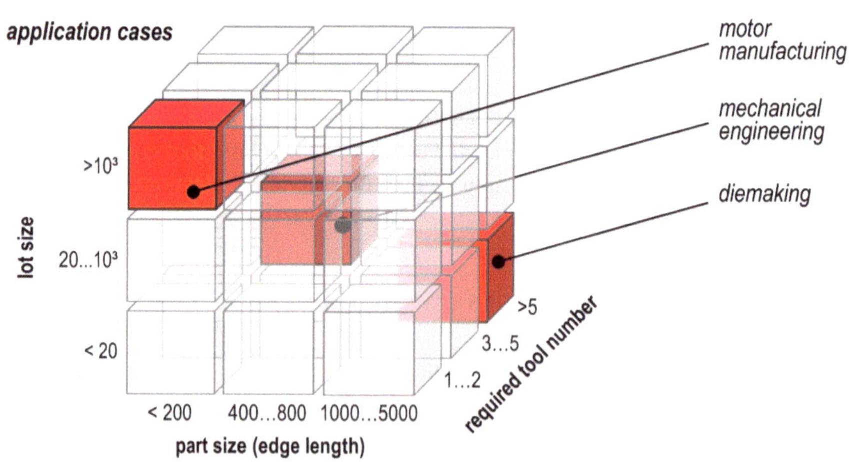

Fig. 20.5 Technological application cases for which the machine tool’s suitability has to be assessed

machine tools are really run. To assess the effect of the methods on the machine tool, the cases of special-purpose machines and series machines are differentiated. This classification makes sense in order to take into account the copy-cost effect in case of a potential multiple use of correction methods.

20.2.3 Benchmarking Criteria

For the reference application cases chosen, the suitability of the compensations and correction methods is judged according to the benchmarking metric in terms of benefit and costs. A method's benefit is benchmarked via properties to be determined, and these properties are described in the following section. When estimating the necessary time and costs, consideration is given to the additional expenses necessary during development, implementation, commissioning and operation of such a machine in comparison with those for a machine tool functioning without compensation and correction methods.

20.2.3.1 Benchmarking Criteria for Benefit Description

The benefit of a method represents the achievable improvements in the machine tool's performance in terms of increased accuracy, shorter machining times and/or lower energy consumption. To determine the parameters for benefit assessment, established methodologies are investigated to determine their usability in general and their efficient usability in particular, as well as, if possible, the potential for modifications designed to meet the specific requirements of the CRC/TR 96. Initially selected evaluation criteria that represent the benefit are detailed.

Machining Accuracy

For a machine tool, the term "machining accuracy" indicates the product quality that can be achieved in the shape and positional tolerances of the part. Different factors, such as geometry, statics and dynamics, as well as the machine's positioning performance, affect the machining accuracy (Verein Deutscher Ingenieure 1977). These factors are not influenced by thermal behaviour and have to be considered separately from the total behaviour for thermal behaviour investigations. For machine tools, this can be done by machine analyses both in the cold and warmed states. The machine is subjected to a thermal load through an appropriate introduction of energy. The changes in geometry that result are either measured (analogously to the standard DIN 1999) or captured through the manufacture of a test part. A functioning test piece was engineered in the subproject.

A method for energy introduction into the machine tool that functions without any additional loading devices was engineered in conjunction with the test piece.

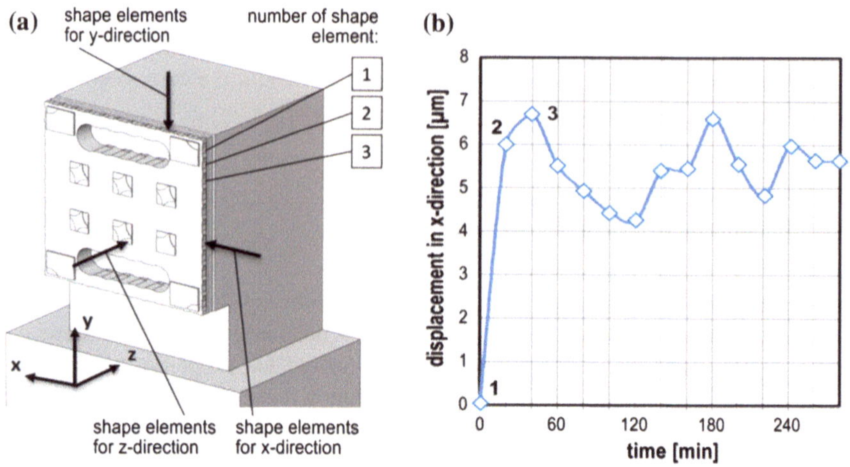

Fig. 20.6 Thermal test piece. **a** Test piece. **b** Thermal displacement

The relevant drives initiate a travel mode, which is carried out cyclically without the machining process. The travel mode has the same energetic effects as a representative machining process. The thermally induced machine tool displacements are transferred in shape elements on the test piece by means of a milling process.

At the beginning of the experiment associated with the test piece, test surfaces are milled in the test piece in the cold machine state. This process is run very quickly and thus virtually no thermally induced machine tool displacements occur during its performance. The thermally induced machine tool displacements generated during the experiment are transferred onto the test piece in defined time intervals. Figure 20.6 shows the test part with all 22 shape elements for the x-, y- and z-directions. Assuming 20 min intervals in the manufacturing sequence of the shape elements, it is possible to execute tests lasting 7 h.

The thermally induced machine tool displacements are found by scanning the test part afterwards in order to finally determine the displacement curves as a function of time depending on direction.

Process Quality (Variance)

Process quality specifies the repeatability with which the machining process can be conducted or the probability values at which special tolerances can be produced on the machine tool (Verein Deutscher Ingenieure 2002). Here only the thermally induced portion of process quality is taken into account. To quantify the process quality, the thermally induced dislocations are first determined for a certain quantity of workpieces by using the methods to investigate machining accuracy. This number is given by the productive machine running time with no breaks, as well as the part's machining time and thus, in turn, the technological application case depending on the operational constraints. Average, standard deviation σ and the process capability index C_P are calculated from the measured displacements of the individual tests (Weck 2006; DIN 1999).

Productivity

A machine tool's productivity can be quantified by the number of workpieces yielded per machine hour. Two tests are required to determine the curve of the thermally induced machine tool displacement as a function of time with different power consumption to warm up the machine tool in order to explore a method's influence on productivity by means of a test piece. It is possible to estimate the influence of increased part output on the maintenance of required tolerances. Based on the results, the number of ok parts and thus the process capability index C_P are derived for two different experimental conditions. Thus, the part output for this nominal value can be calculated based on the specification of a nominal process capability index value C_P of, for instance, 1, in a 6σ process, using the simplified assumption of an inversely proportional relationship between the process capability index C_P and part output. The productivity of these machine tool configurations can be benchmarked by comparing the part output values determined this way for different machine tool configurations.

Energy Consumption

Energy input is captured over the overall test period when determining the curve of the thermally induced machine tool displacements as a function of time. Energy consumption per workpiece can be found by means of the part output (see productivity). This value is used to assess the energy input of the machine tool configuration investigated. Benchmarks of the corresponding results for different machine tool configurations only make sense for the same technological application case (see Fig. 20.5). For an overall consideration of the methods' energetic impacts, for instance, on air conditioning of the factory floor that may not be needed, it is necessary to extend the benchmarking model using the relevant peripheral equipment.

20.2.3.2 Benchmarking Criteria Representing Costs

Expense benchmarking of the individual methods mainly analyses the costs, both in money and time, that occur as a result of the application of a compensation and correction method for each machine tool. It is assumed that the expenditures for the methods differ during the various life cycle phases of a machine (such as development, commissioning, operation). Consequently, a life cycle model for the representation of the relevant realm is used. The expenses (personnel and resource costs) accruing in each phase are analysed by means of process modelling languages [see for instance (Zur Muehlen and Indulska 2010)] in the form of workflow models. These models consider the corresponding circumstances (payload/salary, test capacities, opportunities to carry out simulations, etc.) of the machine tool producer, for which one of the developed methods is benchmarked, on the one hand. On the other hand, incoming information from the subprojects about the expenses to be expected in operation (e.g. for the modification of correction models) is also considered.

Machine Life Cycle Costs

Machine life cycle models represent every phase of the machine's creation, operation and recycling. The VDMA standard sheet no. 34160 presents detailed estimated costs based on a life cycle model (Verband Deutscher Maschinen- und Anlagenbau e.V. 2012). In the context of this project, these estimated costs are modified to identify relevant life cycle phases and are used, rather than merely understood purely as estimates. It is necessary to expand the guidelines for these estimated costs, since they only take into account the operator's view, though the methods also result in expenses for the machine tool manufacturers.

Figure 20.7 illustrates the modified guidelines for cost benchmarking and points out sections for each phase that are influenced by the methods according to the current state of expertise, which are thus analysed. In the "development phase", for instance, additional expenses in machine modelling and simulation result from the methods' application. Thus, it could be necessary, for example, to regularly adjust the correction models in the "operation phase", and these adjustments would

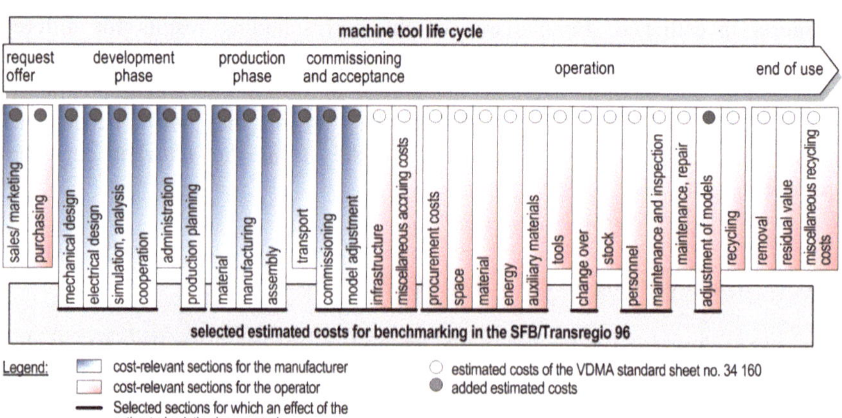

Fig. 20.7 Extended life cycle cost model with explicit focus on the correction or compensation methods

increase the maintenance and inspection costs. The phase "end of use" should not to be considered since it is relatively irrelevant.

From the current perspective, it is expected that additional expenses caused by the correction methods will be caused predominantly by personnel costs. Thus, for instance, experimental or model-based analyses of the thermal machine tool behaviour become necessary. The compensation methods mainly result in increased expenses for the purchase of components and semi-finished products.

The workflows related to the personnel costs are sophisticated and difficult to represent and define in terms of the resources and costs involved. For this reason, the workflows are analysed by means of workflow models.

Engineering Workflows

Modelling of workflows is performed by means of the process modelling language BPMN 2.0 [Business Process Management and Notation, compare (Recker 2008)]. The BPMN was modified for the domain (Braun and Esswein 2014) following an approach by Stroppi et al. (2011) so that resources and temporal or financial aspects of the specific workflows can be represented. A selection of the additional elements' graphic representations is provided in Fig. 20.8. Non-graphic object attributes were also added.

The enquiry for the flow models is performed by interviews conducted regarding the workflows connected with the methods' implementation and the expenses and times spent in the subprojects. The interviews are carried out repeatedly so that the states of progress that have changed due to the research outcomes can be mapped. Experts from industry are asked for their opinions on the solution methods' consequences in practice based on these technology-specific flow models. These interviews are executed by means of the Delphi method to be able to, for instance, estimate expected expenses and required time in the operation phase. The flow

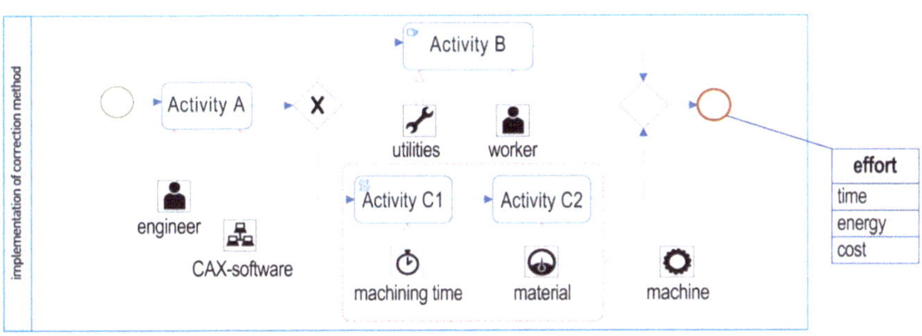

Fig. 20.8 Simplified BPMN model with extension elements (*screenshot* from the Meta modelling tool Cubetto Toolset). From *left* to *right* human resource ("engineer"), application/equipment ("CAX software"), time ("machining time"), auxiliary material, human resource ("worker"), material and machine

models are enhanced by this information. The flow models with the associated resource elements are created for each method (compare Fig. 20.8). In a second step, the parameter values of these models are assigned special values, such as for the number of employee hours required and hourly rates. Assignment of concrete data is done either based on the market prices or on business-individual cost.

It is possible to record the employee hours required per work step by means of the diary method (Rausch et al. 2012) and to estimate the values for the hourly rates or to take them from published wages. Thus it is possible to compare the solutions' methods.

20.3 Model Application

The model for technical-economic benchmarking of the methods to compensate and correct thermally induced machine tool errors is to be applied step by step depending on the research progress of the other subprojects. The analyses in the first phase of the CRC/TR 96 are initially focussed on the engineering-oriented issues of the correction methods. The benchmarking model is applied as mentioned below:

- Representation of the machine tool functions, including the methods applied to the partial model "machine tool configuration",
- Determination of the individual key performance indicators for benchmarking for the machine tool configurations: benefit key performance indicators to be found by means of the thermal test piece; as well as expense key performance indicators by means of workflow modelling,
- Bringing together the individual evaluations by weighed benchmarking of the suitability of the machine tool configurations for each reference case. Weighing makes it possible to consider different priorities resulting from the application cases in the machine properties' analysis. The weighed individual assessments are summarised at the end.

The individual benchmarking key performance indicators are concrete values representing, for instance, the accuracy that can be achieved with one method, related to one application case. Thus they are a means of support for the persons responsible for decision making. The weighed benchmark provides a total evaluation of a method implemented on a machine tool to obtain a rough overview by means of *one* key performance indicator.

20.4 Classification in the CRC/TR 96 and Outlook

In the current project step, the subproject's development flows related to the correction method are analysed so that the expenses incurred in their future industrial implementation can be estimated. Furthermore, these analyses make available expertise related to the different methods' economic and technical consequences for the subprojects for different reference cases. These, in turn, contribute to further enhancement of the methods and to the classification of fields of application in business as a whole.

The plan is to perform benefit analysis for the compensation and correction methods in later project phases, after they have been integrated into demonstrator machines and are thus available for testing.

References

Abele E, Kuhrke B, Rothenbücher S (2010) Entwicklungstrends zur Erhöhung und Bewertung der Energieeffizienz spanender Werkzeugmaschinen. Tagungsband des 1. Internationalen Kolloquiums des Spitzentechnologieclusters eniPROD "Energieeffiziente Produkt- und Prozessinnovationen in der Produktionstechnik"

Braun R, Esswein W (2014) Extending BPMN for modeling resource aspects in the domain of machine tools. WIT Trans Eng Sci 87:450–458

Denkena B, Schürmeyer J, Eikötter M (2010) Berücksichtigung temporärer Effekte von Lebenszykluskosten in der Technologiebewertung. Integration der Kapitalwertmethode in total cost and benefits of ownership. ZWF Zeitschrift für wirtschaftlichen Fabrikbetrieb 105:959–963

DIN Deutsches Institut für Normung e.V. (1999) DIN ISO 230 Prüfregeln für Werkzeugmaschinen

Ehrlenspiel K (2009) Integrierte Produktentwicklung. Denkabläufe, Methodeneinsatz, Zusammenar-beit. Carl Hanser Verlag, München, 2009, 4., überarb. Carl Hanser Verlag, München

Esswein W, Turowski K, Juhrisch M (2010) Modellierung betrieblicher Informationssysteme. Modellgestütztes Management

Eversheim W, Schuh G (2005) Integrierte Produkt- und Prozessgestaltung. Springer, Berlin

Lindner G, Götze U Lebenszykluskostenrechnung als Instrument der energiebezogenen wirtschaftlichen Bilanzierung und Bewertung von Werkzeugmaschinen. In: Neugebauer R, Götze U, Drossel W-G (eds) Methoden der energetisch- wirtschaftlichen Bilanzierung und Bewertung in der Produktionstechnik, pp 115–129

Rausch A, Kögler K, Laireiter A (2012) Tagebuchverfahren zur prozessnahen Datenerhebung in Feldstudien Gestaltungsparameter und Anwendungsempfehlungen. Verl. Empirische Pädagogik, Landau

Recker J-C (2008) BPMN modeling—who, where, how and why. BPTrends 5:1–8

Schlegel A (2002) Konzeption und Einsatzvorbereitung eines Werkzeuges für die Bestimmung der Prozessqualität mittels Kennzahlenüberwachung und wissensbasierter Simulation. Dissertation, TU Chemnitz

Schuette R, Rotthowe T (1998) The guidelines of modeling—an approach to enhance the quality in information models. In: Ling T-W, Ram S, Lee ML (eds) Conceptual modeling—ER. Springer, Berlin, pp 240–254

Stroppi LJR, Chiotti O, Villarreal PD (2011) Extending BPMN 2.0: method and tool support. In: Dijkman R, Hofstetter J, Koehler J (eds) Business process model and notation. Springer, Berlin, pp 59–73

Verband Deutscher Maschinen- und Anlagenbau e.V. (2012) Prognose der Lebenszykluskosten gemäß VDMA Einheitsblatt 34160

Verein Deutscher Ingenieure (1977) Deutsche Gesellschaft für Qualität: VDI/DGQ 3441—Statistische Prüfung der Arbeits- und Positionsgenauigkeit von Werkzeugmaschinen - Grundlagen

Verein Deutscher Ingenieure, Verband der Elektrotechnik Elektronik Informationstechnik (2002) VDI/VDE 3712 Blatt 2—Bestimmung der Maschinenfähigkeit von Dispenssystemen

Weck M (2006) Werkzeugmaschinen 5: Messtechnische Untersuchung und Beurteilung, dynamische Stabilität, 7th edn. Springer, Berlin

Wiemer H, Neidhardt L, Großmann K, Braun R, Esswein W (2014) Reduzierung thermisch bedingter Verlagerungen des TCP von spanenden Werkzeugmaschinen - Verbreitung von Maßnahmen und Nutzen-Aufwands-Bewertung. ZWF Zeitschrift für wirtschaftlichen Fabrikbetrieb 109:225–231

Zor S, Schumm D, Leymann F (2011) A proposal of BPMN extensions for the manufacturing domain. In: Proceedings of 44th CIRP international conference on manufacturing systems

Zur Muehlen M, Indulska M (2010) Modeling languages for business processes and business rules: a representational analysis. Inf Syst 35:379–390. doi:10.1016/j.is.2009.02.006

Chapter 21
Experimental Analysis of the Thermo-Elastic Behaviour of Machine Tools by Means of Selective Thermography and Close-Range Photogrammetry

Knut Großmann, Jens Müller, Marcel Merx and Mirko Riedel

Abstract The measuring methodology developed in subproject C06 provides an efficient analysis of the thermal and thermo-elastic behaviour of machine tools. Contactless measurement of deformations and displacements by means of close-range photogrammetry, as well as temperature measurement using selective thermography make possible high structural resolution with minimal instrumentation and measurement execution effort. In conjunction with distributed data logging, Data acquired inside and outside the control are considered and provide information for a comprehensive database for analysis.

21.1 Introduction

As a rule, temperatures and displacements can only be measured on the surface of structural components. Heat flows can be calculated from kinematic and load parameters by means of power dissipation approaches, assuming known parameter dependencies. Temperatures are mostly acquired in a tactile way on the surface of parts and subassemblies, requiring great effort to fasten, calibrate and wire the temperature sensors. Tactile displacement and deformation measurements demand,

K. Großmann · J. Müller · M. Merx (✉) · M. Riedel
Faculty of Mechanical Engineering, Institute for Machine Tools and Control Engineering, Technical University Dresden, Dresden, Germany
e-mail: marcel.merx@tu-dresden.de

K. Großmann
e-mail: knut.grossmann@tu-dresden.de

J. Müller
e-mail: jens.mueller@tu-dresden.de

M. Riedel
e-mail: Mirko.riedel@tu-dresden.de

K. Großmann (ed.), *Thermo-energetic Design of Machine Tools*,
Lecture Notes in Production Engineering, DOI 10.1007/978-3-319-12625-8_21

as described in Weck (2006), an expensive temperature-stable measuring platform (e.g. a rod framework made of Invar steel). The great instrumentation effort needed for the measurements, as a rule, results in a compromise in the benefit of low structural resolution—in other words, a small number of measuring points. The application of the thermography measurement technique to determine temperatures and temperature fields and the optical measuring method to quantify deformations and displacements offers a method to achieve higher structural resolution. Using the contactless measuring methods of both close-range photogrammetry (Fig. 21.1, left) and selective thermography (Fig. 21.1, middle), it is also possible to significantly reduce the instrumentation effort. In close-range photogrammetry, deformations and displacements are acquired at predefined measuring points on the part's surface. In case of subproject C06 the measuring points are defined by photogrammetry-targets. Selective thermography utilises the information in the visible (optical) range of the spectrum and projects the coordinates of the targets onto the thermal image, thereby making use of the known relative orientation of the optical and thermographic cameras. Temperature is measured in a selective manner at the targets. The targets exhibit a high degree of emissions, which, in turn, enhances the accuracy of the temperature measurement, particularly for shiny metallic surfaces.

Up to now, data available in the machine control, such as revolutions per minute and feed rates, have only been considered to a minimal extent or not at all during the analysis along the thermo-elastic functional chain. But these are useful input data, for instance, for power dissipation models. The database for machine analysis can be significantly expanded by the synchronised acquisition of control-internal and -external measured data by means of distributed data logging (Fig. 21.1, right). Consequently, in the subproject, a software framework was engineered for distributed data logging to be applied to various control systems.

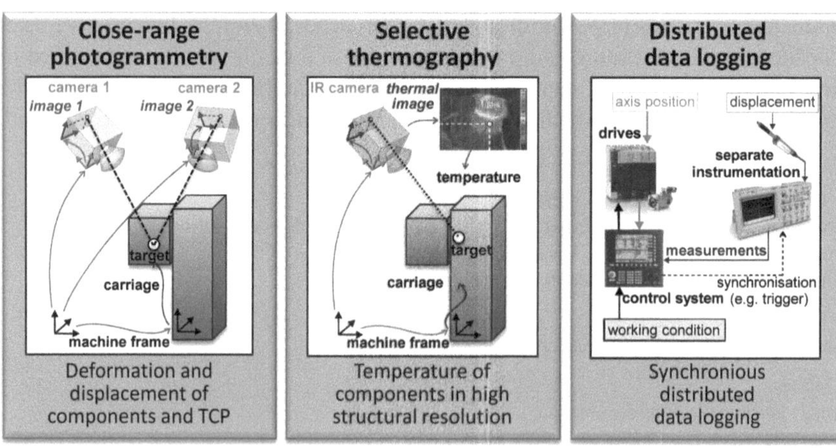

Fig. 21.1 Measuring methodology for efficient analysis of thermo-elastic behaviour according to Großmann (2013a, b)

The measuring methodology according to Fig. 21.1 elaborated in the subproject can be used to analyse complete machine tools and large-sized machine tool sub-assemblies even under working conditions and with minimal instrumentation effort. As a result, the measuring methodology offers a way to generate data for the adjustment of simulation and correction models (for instance, in the subprojects A05, A06, A07, B05, B06, B07), for exemplary parameter identification for components in the mounted-in state (subproject B04) and, finally, to evaluate compensation and correction methods (in the subprojects C05 and C06).

21.2 Approach

21.2.1 Measuring Method of Selective Thermography

The measuring methodology of selective thermography engineered within the scope of the subproject, brings together the advantages of thermographic temperature measurement with those of photogrammetric deformation and displacement measurement [Fig. 21.2, top; comp. Großmann et al. (2013a) and Luhmann (2010)].

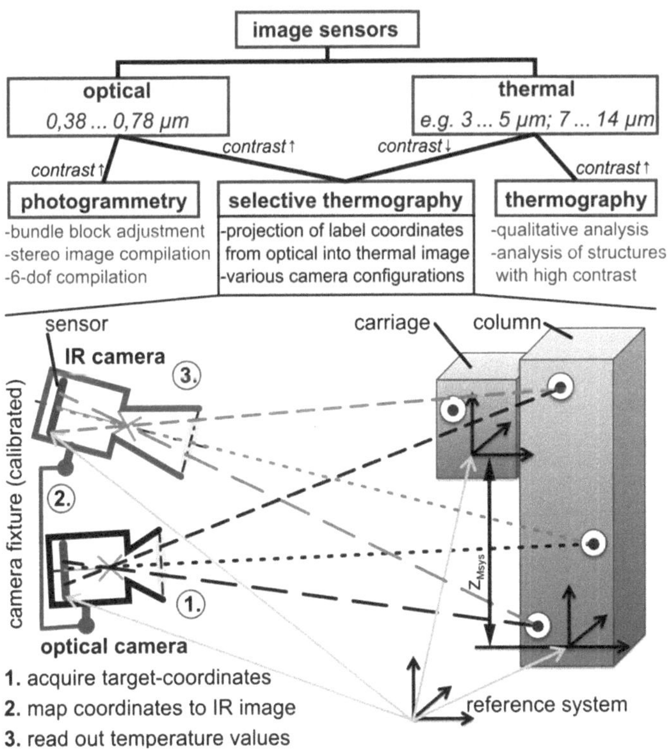

Fig. 21.2 Classification and measurement principle of selective thermography according to Großmann et al. (2013a, b)

It makes possible contactless temperature measurement even in the event of relative motion of object and thermography camera and with insufficient contrast in the thermal image (comp. thermal images in Fig. 21.7). The temperature is measured at defined targets attached to the surface of the object (e.g. printed paper labels). These adhesive photogrammetry targets exhibit a degree of emissions that is known, constant and as high as possible. The spatial location of the targets is determined in the optical image by means of photogrammetric methods (1st in Fig. 21.2, bottom). With a known relative location and orientation of both thermography camera and optical camera in a camera fixture, it is possible to determine where the pixels belonging to the targets lie in the thermal image (2nd in Fig. 21.2, bottom). Temperature is read at these pixels and assigned to the target (3rd in Fig. 21.2, bottom). Comparable approaches were also introduced by Smith (2002) and Acker et al. (2011). However, the latter rely on an areal recording of the object surface in the visible range of the spectrum by means of Speckle Texture Projection Photogrammetry or Strip Light Projection to map the thermal image on a 3D model of the object.

Selective thermography employs a wide range of photogrammetry methods. Accordingly, multiple measurement system configurations and models can be implemented. Three measurement system configurations are shown in an exemplary view in Fig. 21.3.

One can determine the location and orientation of the object to be analysed for each recording period using a 3D model that contains the 3D coordinates of the

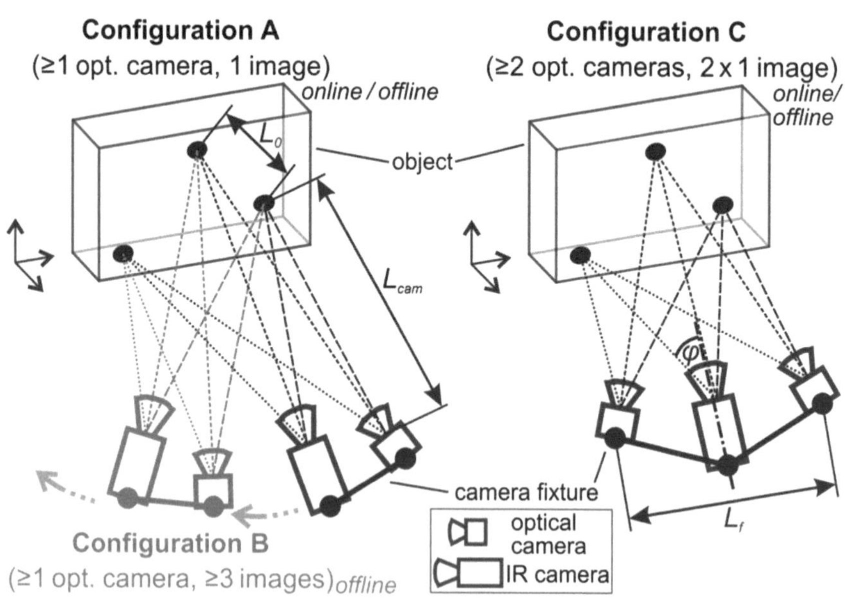

Fig. 21.3 Exemplary measurement system configurations to implement selective thermography according to Großmann et al. (2013a)

targets on the object, and a known scale reference L_0 in the image (configuration A in Fig. 21.3). The 3D coordinates can be projected onto the thermal image given the known relative orientation of optical to thermographic camera inside the camera fixture. For each target, one can also determine angle φ and its distance L_{Cam} from the thermographic camera, thereby enabling the consideration of angle-dependent emission behaviour or of the transmission properties of the medium passed through. Analysis can be conducted online during image acquisition or, as a post-process, after image acquisition, in offline mode. The 3D object model required can be obtained, for example, through preliminary calibration (coordinate measurement) by means of a photogrammetric measurement system.

If in each recording period, at least three images of an object state assumed to be constant are traced from three different perspectives, then the 3D coordinates of the targets can be determined by bundle block adjustment even without knowing the placement of the targets at the object beforehand (configuration B in Fig. 21.3). Adjustment calculation over all periods is only possible after having completed image acquisition, so it is executed offline, as is the projection of the target coordinates in the thermal images.

Having available at least two optical cameras in the camera fixture, assuming a known camera base L_f, it is possible to determine the 3D coordinates of the measured points for each period by triangulation (configuration C in Fig. 21.3). With a measurement system configuration like this, one can measure deformations in each recording period. The coordinates of the targets can also be projected onto the thermal image once after optical image acquisition. Analysis is mostly executed online, since the effort required for calculation is lower than that required for bundle block adjustment.

Both bundle block adjustment (configuration B) and triangulation (configuration C) can be expanded by using a 3D model of the coordinates of the targets attached at the object (comp. configuration A), thus making it possible to find the object's position and orientation.

21.3 Results

21.3.1 Preliminary Investigations and Photogrammetric Deformation and Position Measurements at the Test Bed

The photogrammetric measurement system DPA (Digital Photogrammetric Analysis) by the firm AICON 3D Systems GmbH was used for measurements at the test bed (see Chap. 1) to evaluate the performance of commercially available systems. Figure 21.4, left, illustrates the machine structure equipped with targets, whereas the point cloud recorded is shown in Fig. 21.4, right. When using the manually guided sensor (high resolution digital camera Nikon D3x) for image acquisition from

Fig. 21.4 Photogrammetric measurement by means of AICON DPA at the test bed *left* Photo from the series of images; *right* Recorded point cloud

various perspectives, only insufficient accuracy values could be reached in deformation measurement because of thermal changes during the recording time (approx. 10 … 30 min) and because of varying picture sets. The position of individual moving subassemblies could only be determined with difficulty through the later clustering of the associated measured points in the point cloud and, according to deformation measurement, was too inexact. Brecher (2013) achieved deviations in the range of 10 μm by means of a stationary two-camera-stereo system by the same manufacturer, but with a smaller volume to be measured. However, he only determined the coordinates of the individual measured points, rather than the positions of the moved subassemblies.

The authors' own preliminary tests to evaluate the performance of commercially available systems, and the studies by Brecher (2013) clarify the potential of the photogrammetric recording of thermally affected deformations and displacements on machine tools. The investigations open up issues in research in terms of the optimization of measurement system configurations (number and allocation of cameras) and the extension of the measurement models, for instance, to consider kinematic constraints.

First, selective thermographic temperature measurement, as well as photogrammetric deformation and displacement measurement, were investigated separately at test objects with a simple structure (machine column and Hexapod MiniHex). Two series of measurements based on fundamental studies on the development of software tools and the calibration of measurement systems are introduced in the following chapters.

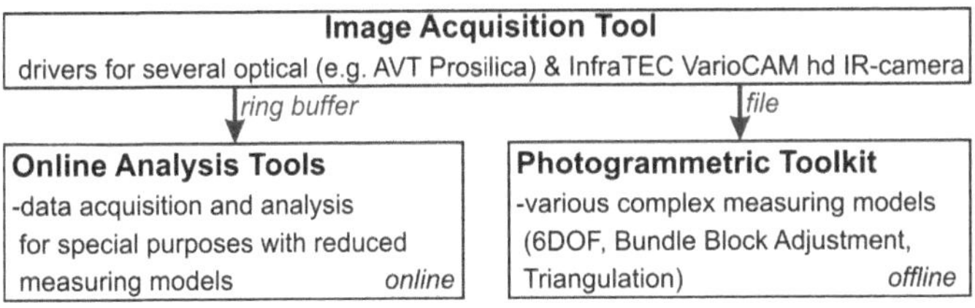

Fig. 21.5 Software tools to execute and analyse photogrammetric and selective thermographic measurements

21.3.2 Development of Software to Execute and Analyse Photogrammetric and Selective Thermographic Measurements

Appropriate software tools were created or refined for the execution and analysis of photogrammetric and selective thermographic measurements. Figure 21.5 gives an overview of the tools applied at the Institute for Machine Tools and Control Technology (IWM). The data recording tool was enhanced by a driver and live image display for the high resolution thermography camera InfraTEC VarioCAM HD.

The photogrammetric toolkit engineered at the IWM and introduced, among others, in Großmann et al. (2012a), was added for selective thermography (Fig. 21.5, right). With the toolkit, camera fixtures can be calibrated and measurements can be analysed by means of extensive measurement models. The toolkit relies on available image data. This means that analysis is conducted offline.

Online analysis tools tailored to the measurement task (Fig. 21.5, left) make it possible to immediately process the image data read from the ring-buffer and to store only the results obtained from it. This strategy makes sense in measurements with less complicated measurement models, but with huge data volumes, for instance, due to the long duration of measurement or the high image acquisition frequency.

21.3.3 Characterisation of Targets and Design of a Procedure to Calibrate a Camera Fixture

Fundamental investigations of the characterisation of selected materials and surface coatings as well of the emittance behaviour of target materials were conducted on a test specimen that was heated in a defined manner, [see Großmann et al. (2013b) for description of test specimen and experimental setup]. The usability of self-sticking paper targets (adhesive photogrammetry targets) for photogrammetric deformation and displacement measurement, as well as thermographic temperature measurement, could be verified. In addition to adhesive targets, tests were also done using targets on magnetic films and active infrared targets (SMD resistors and IR LEDs). However, these are only usable to a limited degree for selective thermography [see Großmann et al. (2013b)].

A technique able to calibrate the camera fixture needed for selective thermography was developed and applied to two camera fixtures [see Großmann et al. (2013b, c)]. The calibration gauge required has to implement "combined" targets visible both in the optical and the thermal images. Calibration delivers sufficiently high accuracy to assign the targets in the thermal image. Calibration inaccuracy values are specified in Großmann et al. (2013d).

21.3.4 Selective Thermographic Temperature Measurement of a Machine Column

The temperature field of the machine column introduced by Großmann et al. (2012c, d) was measured to evaluate the measuring method of selective thermography. Thirteen heating foils attached on each guideway that can be switched individually serve as the thermal load (see Chap. 1 on the experimental setup). Emulated power dissipation corresponds to a spindle head movement of approximately 20 m/min, which is first applied in the bottom range of both guideways (area A in Fig. 21.6) and later in the upper area of the guideway (area B in Fig. 21.6).

The lateral walls and the back side of the machine column are equipped with a total of 20 Pt100 temperature sensors that are attached, as well as six air temperature sensors and two temperature sensors to record the base temperature for commensurable measurement of the temperature field. The selective thermographic temperature measurement is carried out on 17 coded adhesive targets, 7 of those are located at maximal distance of approximately 50 mm to a Pt100 temperature sensor, thus making possible a qualitative comparison of the measured temperature values (see Fig. 21.6 middle and Chap. 1). An AVT Prosilica GX6600 (6,576 × 4,384 pixels) camera was employed as the optical camera. An InfraTec VarioCAM HD (2,048 × 1,536 pixels with resolution enhancement, thermal resolution 0.05 K) was applied to thermography (see Fig. 21.6 right). The measurement system configuration corresponds to the configuration A in Fig. 21.3. However, analysis was conducted after image acquisition (offline) by means of the photogrammetric toolkit. Before starting the test, the machine column equipped with targets was calibrated by means of the photogrammetric measurement system AICON DPA in order to determine the 3D coordinates of the targets and to assign them to the 3D column model.

In Fig. 21.7, the temperatures recorded exemplarily for two selected temperature sensors (T21 left/T26 right column side wall) in a tactile and thermographic manner at next nearest targets are depicted as a function of time. Optical and thermal images with the highlighted target centre points were each mapped for four periods. Additionally the diagram shows the temperature curves calculated for both

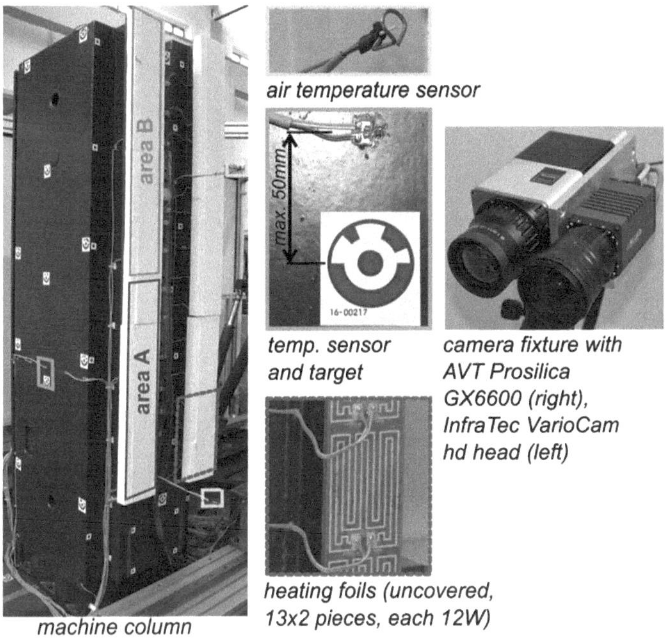

Fig. 21.6 Test setup for machine column analysis by means of selective thermography according to Großmann et al. (2014)

temperature sensors by means of a reduced-order thermal FE model (see Chap. 7) for detail on the MOR-Simulation).

It is also necessary to take into account a slight systematic deviation in the measured values due to the distance of the points measured in a thermographic and tactile manner (see Fig. 21.6 middle).

Additional tests are planned to quantify the achievable measuring inaccuracies.

The measurement described outlines the potential of selective thermography for contactless temperature measurement. It was possible to measure the temperature field in a contactless manner on three sides of the machine column with minimal instrumentation effort (sticking on the adhesive targets and photogrammetric calibration of the targets' coordinates at the structure). A deviation of less than 0.5 °C from the temperature values acquired in a tactile approach was achieved under the existing unfavourable boundary conditions. However, the accuracy potential of the measuring method is far greater.

Adjusting the simulation and measured results (see MOR-Sim in Fig. 21.7), an application was described in which selective thermography—thanks to less instrumentation effort in conjunction with high structural resolution—substantially contributes to the improvement of the machine behaviour analysis along the thermo-elastic functional chain.

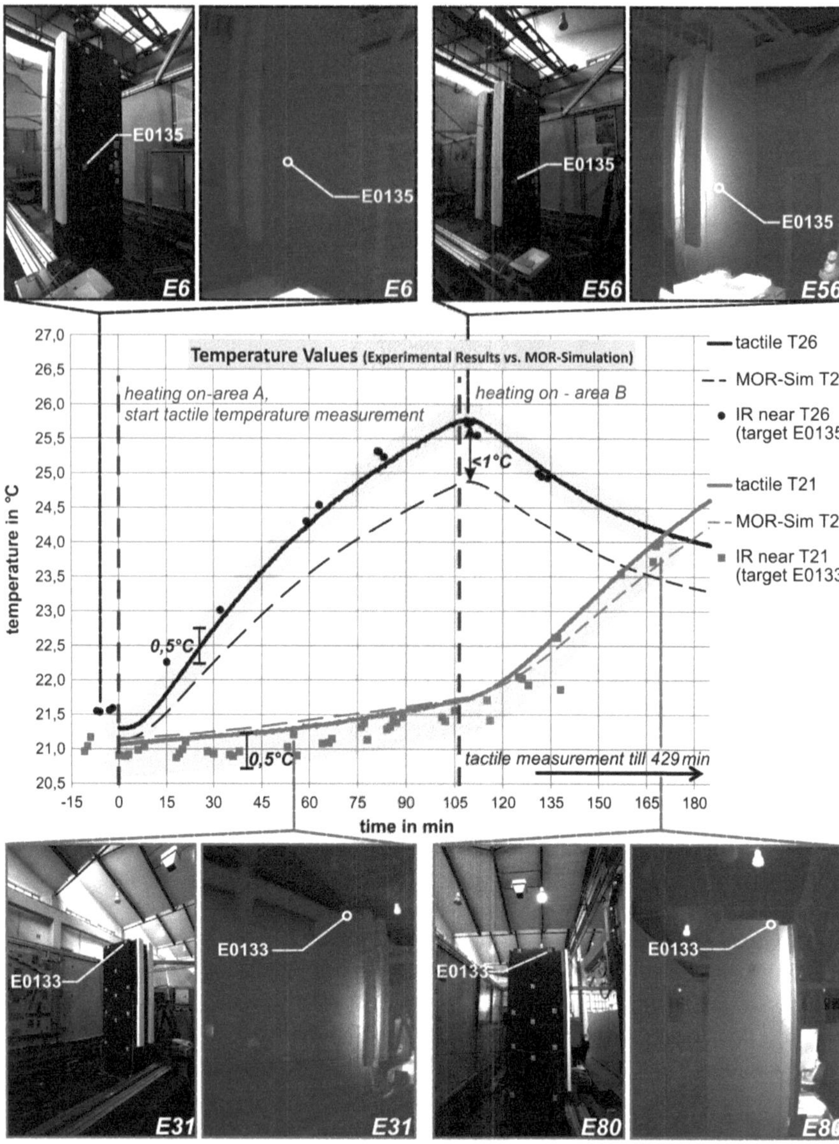

Fig. 21.7 Results of the selective thermographic temperature measurement at machine column according to Großmann (2014)

21.3.5 Photogrammetric Measurement of Thermally Affected Displacements on a Hexapod

Thermally affected displacements were captured by means of a photogrammetric measurement system configuration on the MiniHex (see Chap. 1 and Kauschinger 2006). In Fig. 21.8, left, the test setup given by Großmann et al. (2013c) is shown.

To measure displacements and inclinations, three cameras (AVT Prosilica GC2450 with 2,450 × 2,050 pixels) were attached to the baseplate, and three cameras were affixed to the platform (AVT Manta with 1,280 × 9,64 pixels) (see Fig. 21.8). The cameras on the baseplate have a convergent visual field and thus make it possible to measure the relative platform position (displacement) with high accuracy. The cameras on the platform cover as a group the largest possible visual field (divergent visual field) and thus provide high accuracy in the measurement of the relative angle (inclination) between baseplate and platform. The camera system is self-calibrating, so that 10 various positions were approached for calibration [compare Großmann et al. (2013d)]. Figure 21.8, right, shows the representation of the targets' 3D coordinates and the allocation of cameras after completed calibration. The temperatures of the bearings, ball nuts and the tubes of the rod axes were each measured in a tactile manner and recorded by means of the machine tool control (according to distributed data logging in Fig. 21.1, right). To warm up the machine structure, each rod axis was moved out and retracted again in rapid traverse for 30 times within one load cycle. The traversed path is 250 mm for each axis (see Fig. 21.9, right top). Afterwards, a photogrammetric displacement measurement was carried out automatically in the starting position (see Fig. 21.8, left). These load cycles were repeated 15 times, after which it was switched over to the

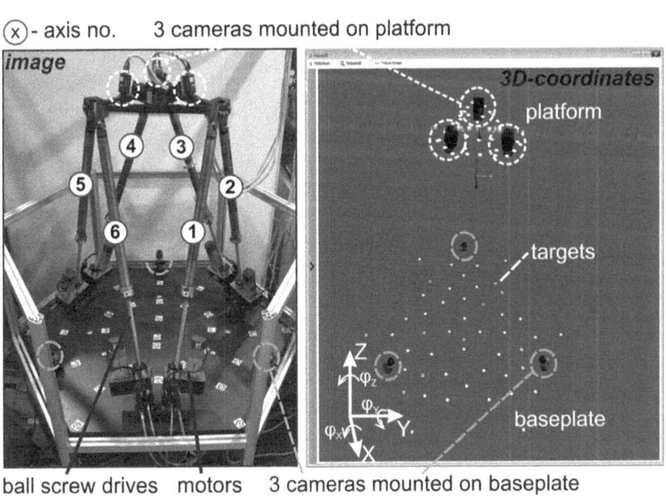

Fig. 21.8 Test setup for the photogrammetric measurement of thermally affected displacements on a hexapod [compare Großmann et al. (2013d)]

next rod. In this way, each rod axis is charged with a load over approximately 12 min and can finally cool down over a period of about 60 min.

Figure 21.9 shows a cutout of the series of measurement lasting altogether approximately 8 h. Figure 21.9, top includes the temperatures measured in a tactile manner (distributed data logging) on the ball nuts of the rod axes. The 15 load cycles on each of the individual rod axes are unambiguously mapped in the temperature curves.

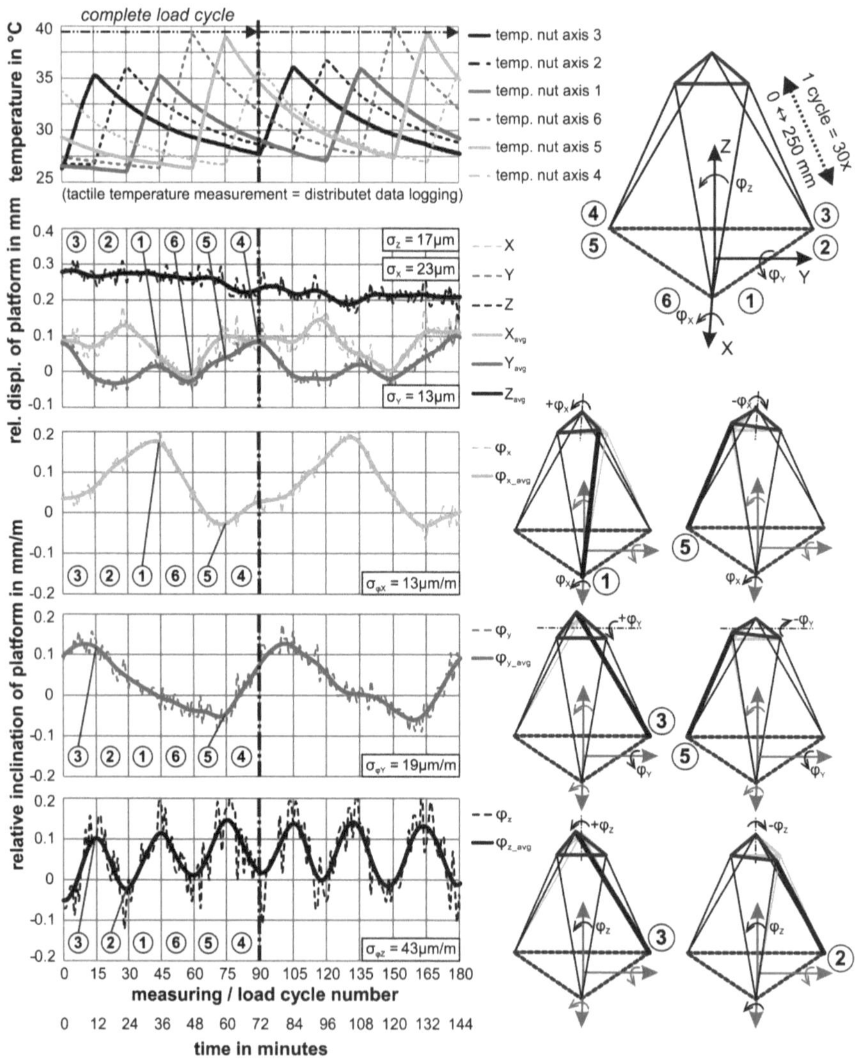

Fig. 21.9 Results of photogrammetric displacement measurement at MiniHex

Whereas the displacement of the platform on the Z axis remains almost constant except for a long-term trend, the load cycles are periodically represented on the X or Y-displacement (see relative displacement of the platform in Fig. 21.9). The displacement in the negative X direction due to warming up of the rod axes 1 and 6 or the displacement (translatory movement) in positive Y direction due to warming up of the rod axes 5 and 4 can be clearly traced.

Thermal expansion of the individual rod axes is obvious in view of torsion of the platform across the Z axis (Fig. 21.9, bottom). Warming of the rod axis 3 results in torsion of the platform in positive φ_Z direction. Warming of the rod axis 2 effects a reverse rotation of the platform in negative φ_Z direction. This sequence is repeated periodically with each change to the next rod axis. Greater periods that correspond to a complete revolution of all six rod axes appear for the inclination of the platform across the X- and Y-axes (see Fig. 21.9, middle).

The results of the series of measurements on the MiniHex (Fig. 21.9) demonstrate the way to measure relative displacements and inclinations on machine tools by means of close-range photogrammetry. There is no problem with, and no limitations on, recording the kinematic deviations in all six degrees of freedom, except for the visibility of targets. Measuring inaccuracy is sufficiently low.

Still lower measuring inaccuracy values can be obtained by an optimisation of the measurement system configuration, which also qualifies the method, as described in Großmann et al. (2012b), also for serial kinematic structures with very low positioning inaccuracy values. The existing series of measurements also clarifies the influence of example-specific variations of the parameters of individual machine components (compare subproject B04). Thus, on the ball nuts of the rod axes 6 and 5, a greater rise in temperature occurs than in the other axes.

21.4 Classification of Outcomes in the CRC/TR 96

In the future, the measurement methodology introduced makes it possible to correlate the temperature field with deformation and displacement, taking into account the operation state of the characterised machine by means of distributed data logging. This allows for a holistic analysis of the machine behaviour along the thermo-elastic functional chain. The measurement methodology primarily benefits those users who have to analyse whole machines under operation conditions “on site”. The methodology provides a major contribution to the adjustment of simulation and correction models (such as in the subprojects A05, A06, A07, B05, B06, B07), for object-related parameter identification, particularly for components in the mounted-in state (subproject B04) and finally the evaluation of compensation and correction methods (subprojects C05 and C06).

The measurement methodology introduced in this paper provides a crucial methodological contribution to the CRC/TR 96.

21.5 Outlook

The measurement methodology according to Fig. 21.1 is used for the comprehensive thermo-elastic analysis of the MiniHex (see Fig. 21.8) and, at the end, implemented at the test bed (see Fig. 21.4) to demonstrate its usability even under poor boundary conditions. Furthermore, additional investigations to quantify the measurement inaccuracy in selective thermographic temperature measurement are planned. The measurement system configurations described in the paper and adequate software tools will be continuously improved and refined.

The techniques to consider the emittance of the targets and the influence of the radiation background are being refined in order to increase the accuracy of the selective thermographic temperature measurement.

In the future, at the test bed, different compensation methods are to be integrated, and correction methods are to be implemented on the PC-based control system by using distributed data logging. The comparison of the methods in terms of their compensation or correction effect is carried out by means of a photogrammetric measurement system configuration installed at the test bed in a fixed manner. Moreover, this measurement system configuration is the basis for a correction approach based on metrology, which will be developed in the third project stage of subproject C06. Within this approach, geometric-kinematic deviations at defined intervals, such as, for instance, in machining idle times are measured photogrammetrically, and correction values for the feed drives are calculated using a kinematic model and corrected.

References

Acker W, Denkena B, Möhring H-C (2011) Berührungslose Erfassungseinrichtung. German Patent DE102009030644B4, 3 Feb 2011

Brecher C, Haber D, Rietdorf A (2013) Thermoelastische Verformung von Werkzeugmaschinen —Thermisch bedingte Strukturverformungen kamerabasiert erfassen und visualisieren. Wt Online 103(5):364–369

Großmann K, Kauschinger B, Riedel M (2012a) Photogrammetrischer Modellbaukasten zur Erfassung von Bewegungsfehlern an Werkzeugmaschinen. In: Photogrammetrie, Laserscanning, optische 3D-Messtechnik—Beiträge der Oldenburger 3D-Tage 2012. Wichmann, Heidelberg

Großmann K, Kauschinger, B, Riedel M (2012b) Photogrammetrische Erfassung von Bewegungsfehlern an Werkzeugmaschinen und Linearachsen. In: 15. Anwendungsbezogener Workshop zur Erfassung, Modellierung, Verarbeitung und Auswertung von 3D-Daten. Gesellschaft zur Förderung angewandter Informatik e.V. (GFaI), Berlin

Großmann K, Städel C, Galant A, Mühl A (2012c) Berechnung von Temperaturfeldern an Werkzeugmaschinen—Vergleichende Untersuchung alternativer Methoden zur Erzeugung kompakter Modelle. ZWF 107(6):452–456

Großmann K, Galant A, Mühl A (2012d) Effiziente Simulation durch Modell-ordnungs-reduktion—Thermo-elastische Berechnung von Werkzeug-maschinen-Baugruppen. ZWF 107(6):457–461

Großmann K, Müller J, Merx M, Riedel M (2013a) Untersuchung des thermo-elastischen Verhaltens von Werkzeugmaschinen—Grundlagen der experimentellen Analyse mit Hilfe der selektiven Thermografie. ZWF 108(7–8):492–497

Großmann K, Merx M, Riedel M (2013b) Thermografie und Nahbereichsphotogrammetrie zur Erfassung von Temperatur- und Verlagerungsfeldern. In: Tradition und Gegenwart bei der Analyse des thermischen Verhaltens spanender Werkzeugmaschinen, DWM e.V., Dresden

Großmann K, Merx M, Riedel M (2013c) Temperatur- und Verlagerungsmessung an einem Versuchsträger mittels selektiver Thermografie und Nahbereichs-Photogrammetrie. Paper presented at the 3rd CRC/TR colloquium, RWTH Aachen, Aachen, 29–30 Oct 2013. http://nbn-resolving.de/urn:nbn:de:bsz:14-qucosa-143947

Großmann K, Riedel M, Merx M (2013d) Thermografie und Nahbereichsphotogrammetrie zur Erfassung von Temperatur- und Verlagerungsfeldern. In: 16. Anwendungs-bezogener Workshop zur Erfassung, Modellierung, Verarbeitung und Auswertung von 3D-Daten. Gesellschaft zur Förderung angewandter Informatik e.V. (GFaI), Berlin, ISBN 978-3-942709-09-5

Großmann K, Galant A, Merx M, Riedel M (2014) Verfahren zur effizienten Analyse des thermoelastischen Verhaltens von Werkzeugmaschinen. In: Neugebauer R, Drossel W-G (eds) Innovations of sustainable production for green mobility—energy-efficient technologies in production, vol 80. Reports from the IWU, pp 683–699

Kauschinger B (2006) Verbesserung der Bewegungsgenauigkeit an einem Hexapod einfacher Bauart. Dissertation, Technische Universität Dresden

Luhmann T (2010) Nahbereichsphotogrammetrie—Grundlagen, Methoden und Anwendungen. Wichmann, Berlin

Smith ST (2002) Modelling hot bodies—combined real-time 3D and thermal imaging for medical applications. 3D-MATIC department, Glasgow University

Weck M, Brecher C (2006) Werkzeugmaschinen—Messtechnische Untersuchung und Beurteilung, dynamische Stabilität. Springer, Berlin

MIX
Papier aus verantwortungsvollen Quellen
Paper from responsible sources
FSC® C105338

If you have any concerns about our products,
you can contact us on
ProductSafety@springernature.com

In case Publisher is established outside the EU,
the EU authorized representative is:
Springer Nature Customer Service Center GmbH
Europaplatz 3, 69115 Heidelberg, Germany

Printed by Libri Plureos GmbH
in Hamburg, Germany